TRANSITIONAL MATHEMATICS

Developing Number Sense

SOLUTIONS BOOK

John Woodward, Ph.D. and Mary Stroh, B.S.

Sopris West Educational Services • Longmont, Colorado

ISBN 1-57035-962-8

Edited by Beverly Rokes
Editorial assistance by Annette Reaves
Text design by Edward Horcharik *and* Sebastian Pallini
Text production by Edward Horcharik *and* Matthew Williams
Cover design by Sue Campbell
Production assistance by Denise Geddis

06 05 04 03 6 5 4 3 2 1

Printed in the United States of America

Published and Distributed by

SOPRIS WEST
EDUCATIONAL SERVICES

4093 Specialty Place • Longmont, Colorado 80504 • (800) 547-6747
www.sopriswest.com

212SOLU/11-03/BAN/4M/089

UNIT 1

STUDENT TEXTBOOK ANSWER KEY
Addition

Lesson 1—Homework
PART 1
1. 2,000,000
2. 7
3. 90,000
4. 0
5. 5

PART 2
1. 10
2. 90
3. 20,000,000
4. 300
5. 20,000
6. 100,000
7. 0
8. 3,000,000
9. 2,000
10. 7,000
11. 500,000,000
12. 70

PART 3
1. 65,029
2. 74,160
3. 813
4. 4,000,025

Lesson 2—Homework
PART 1 (Note: The zeros are optional placeholders.)
1. 70 + 5
2. 900 + 0 + 7
3. 200 + 90 + 0
4. 400 + 70 + 8
5. 500 + 50 + 5
6. 1,000 + 600 + 30 + 9
7. 300 + 0 + 5
8. 6,000 + 0 + 50 + 3
9. 200 + 10 + 3
10. 900 + 90 + 9
11. 800 + 50 + 0
12. 10,000 + 0 + 0 + 0 + 1

PART 2
1. 89
2. 406
3. 580
4. 662
5. 999
6. 1,009
7. 3,011
8. 427
9. 810
10. 66
11. 4,200
12. 777

PART 3
1. 9,000,000
2. hundred thousands place
   ten thousands place
   tens place
3. 0
4. 7
5. 2

Lesson 3—Homework
PART 1
1. 16
2. 12
3. 18
4. 15
5. 10
6. 10
7. 15
8. 13
9. 13
10. 14
11. 11
12. 11
13. 17
14. 14
15. 10
16. 16

PART 2
1. 12
2. 120
3. 1,200
4. 15
5. 150
6. 1,500
7. 15
8. 150
9. 1,500
10. 14
11. 140
12. 1,400

PART 3
1. 74
2. 709
3. 505
4. 9,131
5. 799
6. 8,105

PART 4
1. 600 + 50 + 2
2. 500 + 80 + 8
3. 20 + 9
4. 5,000 + 800 + 70 + 7
5. 600 + 30 + 0
6. 1,000 + 800 + 0 + 2

PART 5
1. The problem is asking: How many goals did Josh and Michael score altogether?

2. The problem is asking: How many books did the family read?

Lesson 4—Homework

PART 1

1.
```
  35
+ 28
```
2.
```
  47
+ 65
```
3.
```
  18
+ 14
```
4.
```
 300
+ 20
```
5.
```
  20
+ 30
```
6.
```
 501
+ 22
```
7.
```
  612
+ 408
```
8.
```
  207
+ 390
```
9.
```
  620
+ 330
```

PART 2

1. 98 + 12
2. 46 + 34
3. 18 + 172

PART 3 (Note: The zeros are optional placeholders.)

1. 30 + 5
2. 500 + 20 + 9
3. 800 + 10 + 2
4. 300 + 70 + 5
5. 10,000 + 6,000 + 0 + 20 + 0
6. 40,000 + 5,000 + 900 + 90 + 9
7. 6,000 + 0 + 10 + 5
8. 4,000 + 0 + 0 + 7
9. 20,000 + 0 + 600 + 0 + 1

PART 4

1. 7
2. 500
3. 10
4. 90
5. 8,000,000
6. 2
7. 6,000
8. 0
9. 30,000

Lesson 5—Homework

PART 1

1.

	30	5
+	20	8

2.

	100	0	7
+		80	0

3.

	40	7
+	60	5

4.

	400	90	5
+	100	60	2

5.

	200	0	8
+		50	1

6.

	300	90	0
+	900	20	0

7.

	500	60	0
+		70	2

8.

	300	80	2
+	100	10	2

PART 2

1.
```
  38
+ 46
```
2.
```
  29
+ 75
```
3.
```
  50
+ 11
```
4.
```
 538
+ 46
```
5.
```
  709
+ 600
```
6.
```
  998
+ 801
```

PART 3

1. 16
2. 160
3. 1,600
4. 13
5. 130
6. 1,300
7. 12
8. 120
9. 1,200

PART 4

1. The problem is asking: How many times does the human heart beat in an hour?
2. The problem is asking: How much does the world's population grow in a week?

Lesson 6—Homework

PART 1

1.

20	2
+ 30	9

2.

40	3
+ 60	6

3.

10	2
+ 50	9

4.

60	1
+ 20	9

5.

30	0
+ 40	0

6.

10	0
+ 50	9

7.

90	7
+ 10	1

8.

20	0
+ 50	0

9.

60	9
+ 80	7

PART 2

1.

10	7
+ 80	2
90	9

99

2.

40	5
+ 30	3
70	8

78

3.

50	4
+ 20	2
70	6

76

4.

60	1
+ 10	6
70	7

77

5.

30	2
+ 20	5
50	7

57

6.

40	0
+ 40	9
80	9

89

7.

20	2
+ 10	1
30	3

33

8.

50	0
+ 20	0
70	0

70

9.

60	1
+ 30	3
90	4

94

PART 3 (Items in bold type are given in the text.)

Ones	Tens	Hundreds
2 + 4	**20 + 40**	**200 + 400**
2 + 8	**20 + 80**	200 + 800
9 + 1	90 + 10	900 + 100
3 + 6	30 + 60	**300 + 600**
4 + 5	40 + 50	**400 + 500**
2 + 7	20 + 70	200 + 700
3 + 3	**30 + 30**	300 + 300
9 + 8	90 + 80	900 + 800
6 + 5	60 + 50	**600 + 500**
5 + 2	**50 + 20**	500 + 200

Lesson 7—Reading a Bar Graph

1. Number of CDs Sold
2. Month
3. Kazoom Record Company, "The Blockheads," CDs Sold Jan-April 1999
4. 100s
5. Jan—100
   Feb—200
   Mar—300
   April—400

Lesson 7—Homework

PART 1

1.

100	10	7
+	80	2
100	90	9

199

2.

400	40	5
+	30	3
400	70	8

478

3.

900	50	4
+	20	2
900	70	6

976

4.

100	60	1
+	20	2
100	80	3

183

5.

300	30	2
+ 400	20	5
700	50	7

757

6.

500	40	0
+ 100	40	9
600	80	9

689

7.

100	0	2
+ 300	10	1
400	10	3

413

8.

100	50	0
+ 500	20	0
600	70	0

670

9.

100	60	1
+ 200	0	3
300	60	4

364

PART 2

1. The names of the bands are different. The first graph talks about the Blockheads. The second graph talks about the Toasters.
2. The scale for the first graph is 100s. It is counting by 100. The scale for the second graph is 200s. It's counting by 200.

PART 3

1. 400 + 20 = 420
2. 200 + 5 = 205
3. 600 + 50 = 650
4. 100 + 10 = 110
5. 50 + 3 = 53

Lesson 8—Answering Questions About A Graph

1. The difference in sales between January and February is 100.
   The extended fact is 100 + 100 = 200.
2. The difference in sales between February and March is 100.
   The extended fact is 200 + 100 = 300.
3. The increase in sales between April and May is 200.
   The extended fact is 400 + 200 = 600.
   The trend is: Sales are increasing (going up).
   Sample answers about what may be happening to make CD sales go up:
   - The CDs are getting more playing time on the radio.
   - The company is advertising the CDs more.
4. The trend is: Sales are decreasing (going down).
   Sample answers about what may be happening to make CD sales go down:
   - Another CD may have been released that is more popular.
   - The CD has been out for a while and people are tired of the songs.

Lesson 8—Homework

PART 1

1. 16
2. 160
3. 1,600
4. 16
5. 10
6. 100
7. 1,000
8. 10
9. 130
10. 1,300
11. 13
12. 13
13. 17
14. 17
15. 170
16. 1,700

PART 2

1.

200	20	8
+	10	1
200	30	9

239

2.

500	30	4
+	20	3
500	50	7

557

3.

900	50	4
+	30	5
900	80	9

989

4.

200	10	6
+ 100	20	2
300	30	8

338

5.

400	30	2
+ 300	20	5
700	50	7

757

6.

800	0	2
+ 100	0	3
900	0	5

905

7.

100	0	2
+ 100	90	0
200	90	2

292

8.

400	20	0
+ 400	20	0
800	40	0

840

9.

100	0	2
+ 600	0	3
700	0	5

705

PART 3

1.

300	20	8
+ 200	10	1
500	30	9

539

2.

300	0	3
+	30	1
300	30	4

334

PART 4

1. The difference is 200. The extended fact is 200 + 200 = 400.
2. The difference is 600. The extended fact is 200 + 600 = 800.
3. They are alike because their sales are increasing. They are different because the Toasters' sales are going up twice as fast as the Blockheads' sales. The Blockheads' sales are going up by 100 per month and the Toasters' sales are going up 200 per month.

Lesson 9—Homework

PART 1

1. 9
2. 160
3. 900
4. 15
5. 70
6. 700
7. 130
8. 700
9. 7

PART 2

1.

300	30	7
+	20	1
300	50	8

358

2.

200	20	0
+ 100	0	0
300	20	0

320

3.

500	0	0
+ 100	0	2
600	0	2

602

4.

300	0	2
+ 500	0	7
800	0	9

809

5.

400	50	5
+ 200	40	0
600	90	5

695

6.

800	70	1
+ 300	0	0
1,100	70	1

1,171

PART 3

1. May: 400 + 100 = 500
   June: 500 + 100 = 600
   July: 600 + 100 = 700
   Answer: 700
2. You might make the scale smaller; for instance, 50.
3. Yes, just make the bars go halfway between. For instance, if the number is 150, make the bar halfway between 100 and 200.

Lesson 10—Creating a Graph From a Table of Data

Sample Answer:

Kazoom Record Company
The Hammerheads
CDs Sold Jan–April 2000

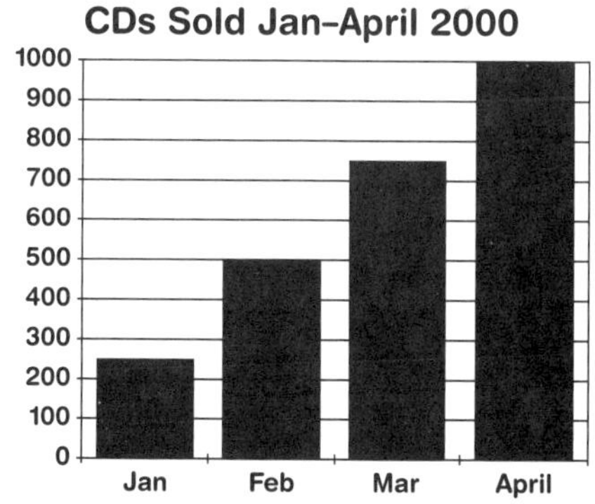

Lesson 10—Homework

PART 1

1.

300	30	7
+	20	1
300	50	8

358

2.

800	60	7
+	10	2
800	70	9

879

3.

600	20	5
+	50	0
600	70	5

675

4.

300	0	2
+ 500	0	7
800	0	9

809

5.

100	20	0
+ 200	40	0
300	60	0

360

6.

800	10	2
+ 100	70	3
900	80	5

985

PART 2

1. $\begin{array}{r} \scriptstyle 1 \\ 37 \\ +\ 24 \\ \hline 61 \end{array}$

2. $\begin{array}{r} 20 \\ +\ 80 \\ \hline 100 \end{array}$

3. $\begin{array}{r} \scriptstyle 1 \\ 79 \\ +\ 13 \\ \hline 92 \end{array}$

4. $\begin{array}{r} \scriptstyle 1 \\ 31 \\ +\ 89 \\ \hline 120 \end{array}$

5. $\begin{array}{r} \scriptstyle 1 \\ 95 \\ +\ 27 \\ \hline 122 \end{array}$

6. $\begin{array}{r} \scriptstyle 1 \\ 87 \\ +\ 25 \\ \hline 112 \end{array}$

PART 3

Sample Answer:

Kazoom Record Company
Third Ear Deaf
Yearly CD Sales 1987–1991

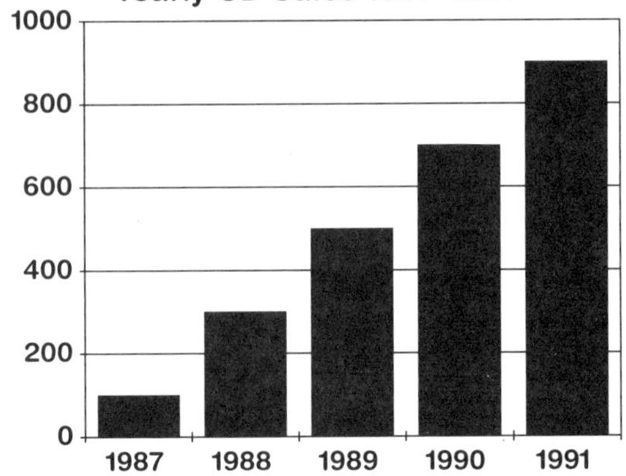

PART 4

1. The increase in sales from 1987 to 1991 was 800. The extended fact is 100 + 800 = 900.
2. The sales increase from 1987 to 1988 was 200. The extended fact is 100 + 200 = 300. The sales increase from 1988 to 1989 was 200. The extended fact is 300 + 200 = 500.
3. The sales are increasing by about 200 CDs per year.

Lesson 11—Homework

PART 1

1. 93
2. 60
3. 81
4. 56
5. 74
6. 80
7. 76
8. 70
9. 80

PART 2

The methods are the same in the following ways:

- They both require regrouping.
- They both result in the same answer.
- The both show a "carry."

The methods are different in the following ways:

- One takes four steps; the other takes one step.
- One shows a "carry" of 10; the other shows a "carry" of 1.

PART 3 (A range of acceptable answers is given for each.)

1. 87 or 88
2. 4, 5, or 6
3. 260–290
4. 505–520
5. 110–140
6. 50–70

Lesson 12—Homework

PART 1

1. 393
2. 460
3. 581
4. 656
5. 324
6. 540
7. 131
8. 1,010
9. 920

PART 2

The methods are the same in the following ways:

- They both require regrouping.
- They both result in the same answer.
- They both show a "carry."

The methods are different in the following ways:

- One takes four steps; the other takes one step.
- One shows a "carry" of 10; the other shows a "carry" of 1.
- One shows a "carry" of 100; the other shows a "carry" of 1.

PART 3

1.
```
  361
+ 142
  503
```

2.
```
  443
+ 239
  682
```

3.
```
  732
+  99
  831
```

PART 4 (A range of acceptable answers is given for each question.)

1. 31 or 32
2. 84 or 85
3. 1 or 2
4. 630–640
5. 905–920

Lesson 13—Rounding Sales Figures to Make a Graph

Sample answers (Students may choose a different scale when rounding.):

Rounded Sales Figures:

Jan.: 200
Feb.: 400
Mar.: 600
Apr.: 800

Sample bar graph:

Kazoom Record Company
The Moon Bandits
CDs Sold Jan–April 2001

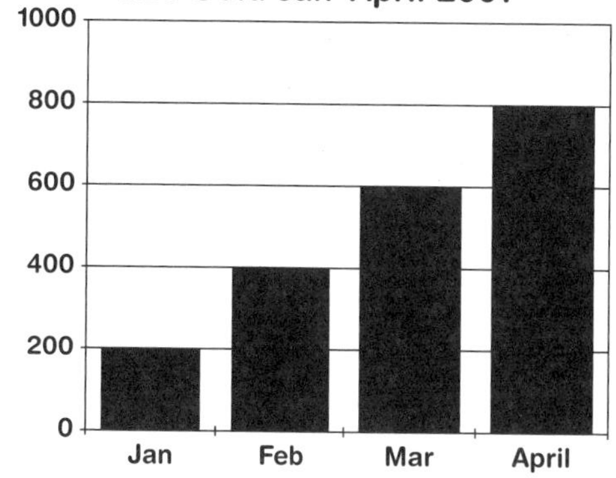

Lesson 13—Homework

PART 1

1. 15
2. 150
3. 1,500
4. 15
5. 15
6. 150
7. 1,500
8. 13
9. 130
10. 1,300
11. 13
12. 13
13. 13
14. 130
15. 130
16. 1,300

PART 2 (Check student work for correct use of number lines.)

1. Estimate: 70 + 80 = 150
   Exact Answer: 150
2. Estimate: 60 + 70 = 130
   Exact Answer: 132
3. Estimate: 50 + 80 = 130
   Exact Answer: 138
4. Estimate: 90 + 60 = 150
   Exact Answer: 153
5. Estimate: 60 + 90 = 150
   Exact Answer: 153
6. Estimate: 50 + 80 = 130
   Exact Answer: 136
7. Estimate: 500 + 800 = 1,300
   Exact Answer: 1,298
8. Estimate: 600 + 900 = 1,500
   Exact Answer: 1,526
9. Estimate: 500 + 800 = 1,300
   Exact Answer: 1,310
10. Estimate: 700 + 600 = 1,300
    Exact Answer: 1,318
11. Estimate: 600 + 700 = 1,300
    Exact Answer: 1,322
12. Estimate: 700 + 800 = 1,500
    Exact Answer: 1,527

PART 3

1. 787 is about 800 and 445 is about 400.
   800 + 400 = 1,200.
   They have only about 1,200 tickets. They need 1,500. They do not have enough tickets to get the prize.
2. 227 is about 200, 329 is about 300, and 179 is about 200. 200 + 300 + 200 = 700. They traveled about 700 miles.

Lesson 14—Horizontal Bar Graphs

1. Innex Records sold about 350 records.
2. The combined sales of Kazoom and Recordex was about 1,300 (600 + 700 = 1,300).
3. BMZ Inc. would have to sell about 100 more records to match Recordex's sales (600 + 100 = 700).

Lesson 14—Drawing Horizontal Bar Graphs

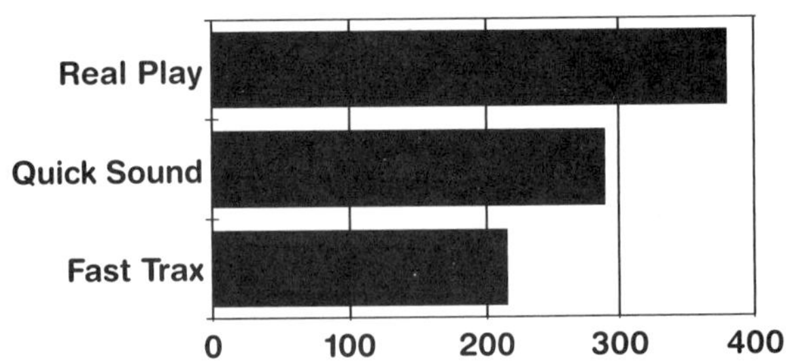

Lesson 14—Homework

PART 1

1. 8
2. 150
3. 800
4. 16
5. 70
6. 700
7. 110
8. 400
9. 4

PART 2

1. 72 is about 70, and 81 is about 80.
   70 + 80 = 150
   Actual Answer: 153
2. 68 is about 70, and 94 is about 90.
   70 + 90 = 160
   Actual Answer: 162
3. 44 is about 40, and 74 is about 70.
   40 + 70 = 110
   Actual Answer: 118
4. 94 is about 90, and 67 is about 70.
   90 + 70 = 160
   Actual Answer: 161
5. 39 is about 40, and 74 is about 70.
   40 + 70 = 110
   Actual Answer: 113
6. 84 is about 80, and 72 is about 70.
   80 + 70 = 150
   Actual Answer: 156
7. 399 is about 400, and 701 is about 700.
   400 + 700 = 1,100
   Actual Answer: 1,100
8. 789 is about 800, and 737 is about 700.
   800 + 700 = 1,500
   Actual Answer: 1,526
9. 411 is about 400, and 699 is about 700.
   400 + 700 = 1,100
   Actual Answer: 1,110
10. 839 is about 800, and 679 is about 700.
    800 + 700 = 1,500
    Actual Answer: 1,518
11. 701 is about 700, and 921 is about 900.
    700 + 900 = 1,600
    Actual Answer: 1,622
12. 442 is about 400, and 715 is about 700.
    400 + 700 = 1,100
    Actual Answer: 1,157

PART 3

1. The answer should be 106.
2. The answer is correct.
3. The answer should be 210.
4. The answer should be 68.
5. The answer is correct.
6. The answer should be 101.

PART 4

1. 429 is about 400, and 389 is about 400.
   400 + 400 = 800
   There were about 800 students at the two events.
   Actual Answer: 818 students
2. 845 is about 800, and 487 is about 500.
   800 + 500 = 1,300
   Bob spent about $1,300 on the car and the parts.
   Actual Answer: $1,332
3. 398 is about 400, and 321 is about 300.
   400 + 300 = 700.

Stephanie's new house is about 700 miles away.
Actual Answer: 719 miles.

Lesson 15—Homework

PART 1

1. 7
2. 70
3. 700
4. 17
5. 170
6. 1,700
7. 110
8. 1,100
9. 11

PART 2

1. 12 ⟶ 10 + 2
2. 35 ⟶ 30 + 5
3. 49 ⟶ 40 + 9
4. 60 ⟶ 60 + 0
5. 500 ⟶ 500 + 0 + 0
6. 602 ⟶ 600 + 0 + 2
7. 812 ⟶ 800 + 10 + 2
8. 975 ⟶ 900 + 70 + 5
9. 999 ⟶ 900 + 90 + 9
10. 537 ⟶ 500 + 30 + 7
11. 4,092 ⟶ 4,000 + 0 + 90 + 2
12. 6,109 ⟶ 6,000 + 100 + 0 + 9

PART 3

1.

	70	2
+	10	5
	80	7

87

2.

	30	2
+	20	3
	50	5

55

3.

	30	1
+	10	7
	40	8

48

4.

	40	3
+	20	6
	60	9

69

5.

	40	0
+	20	7
	60	7

67

6.

	20	7
+	40	2
	60	9

69

PART 4

1. 94
2. 61
3. 55
4. 72
5. 76
6. 72

Lesson 16—More Horizontal Bar Charts

1. The most common length seems to be 111–120 minutes.
2. About 49 (7 + 18 + 24) movies were between 91 and 120 minutes in length.
3. Harry looked at about 57 movies all together (5 + 7 + 18 + 24 + 3).
4. Sample Answer (vertical bar chart)

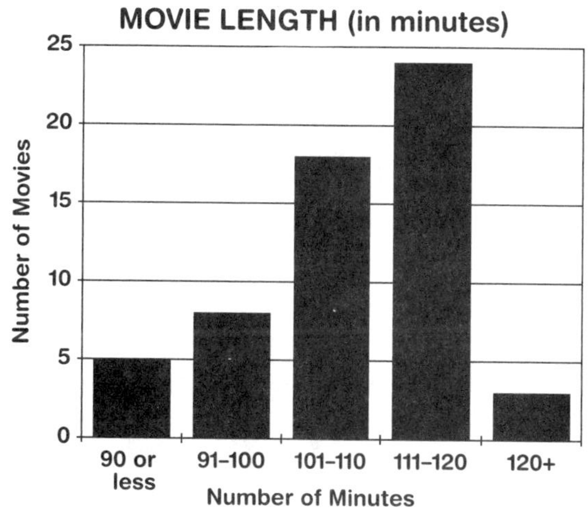

Lesson 16—Homework

PART 1

1. 20 + 2 + 30 + 5 = 20 + 30 + 2 + 5 = 50 + 7 = 57
2. 40 + 3 + 30 + 6 = 40 + 30 + 3 + 6 = 70 + 9 = 79
3. 10 + 2 + 50 + 6 = 10 + 50 + 2 + 6 = 60 + 8 = 68
4. 70 + 1 + 10 + 3 = 70 + 10 + 1 + 3 = 80 + 4 = 84
5. 20 + 5 + 40 + 0 = 20 + 40 + 5 + 0 = 60 + 5 = 65
6. 30 + 0 + 50 + 9 = 30 + 50 + 0 + 9 = 80 + 9 = 89

PART 2

1. 30 + 20 + 4 + 1 = 50 + 5 = 55
   34 + 21 = 55
2. 20 + 10 + 3 + 5 = 30 + 8 = 38
   23 + 15 = 38
3. 70 + 10 + 6 + 0 = 80 + 6 = 86
   76 + 10 = 86
4. 60 + 30 + 0 + 2 = 90 + 2 = 92
   60 + 32 = 92
5. 10 + 80 + 1 + 8 = 90 + 9 = 99
   11 + 88 = 99
6. 50 + 20 + 7 + 0 = 70 + 7 = 77
   57 + 20 = 77

PART 3

1. 66
2. 153
3. 111
4. 226
5. 172
6. 586

Lesson 17—Creating a Problem From Data

Sample Answer:

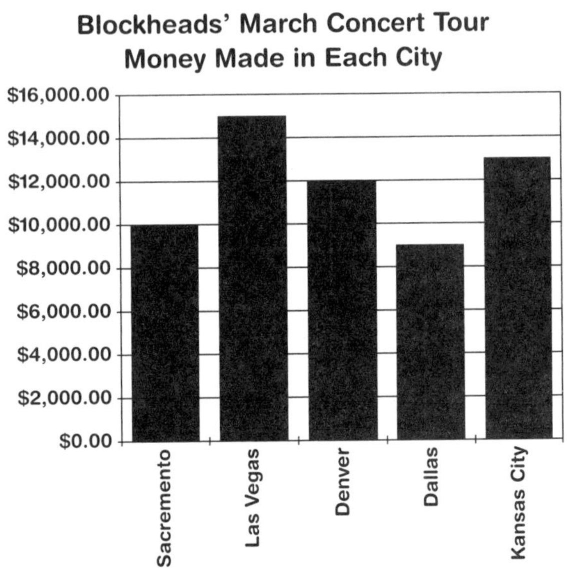

Sample Problems:

1. How much money did the Blockheads make all together in all of the cities?
   Answer: $10,000 + $15,000 + $12,000 + $9,000 + $13,000 = $59,000
2. Which concert paid the most?
   Answer: Las Vegas
3. Which concert paid the least? Answer: Dallas

Lesson 17—Homework

PART 1

1. 30 + 7 + 10 + 1 = 30 + 10 + 7 + 1=
   40 + 8 = 48
2. 20 + 6 + 60 + 2 = 20 + 60 + 6 + 2 =
   80 + 8 = 88
3. 30 + 8 + 40 + 0 = 30 + 40 + 8 + 0 =
   70 + 8 = 78
4. 10 + 5 + 30 + 1 = 10 + 30 + 5 + 1 =
   40 + 6 = 46
5. 20 + 2 + 30 + 3 = 20 + 30 + 2 + 3 =
   50 + 5 = 55
6. 40 + 2 + 50 + 3 = 40 + 50 + 2 + 3 =
   90 + 5 = 95
7. 80 + 0 + 10 + 0 = 80 + 10 + 0 + 0 =
   90 + 0 = 90
8. 10 + 3 + 20 + 4 = 10 + 20 + 3 + 4 =
   30 + 7 = 37
9. 40 + 6 + 20 + 1 = 40 + 20 + 6 + 1 =
   60 + 7 = 67
10. 60 + 6 + 10 + 2 = 60 + 10 + 6 + 2 =
    70 + 8 = 78

PART 2

1. 137 is about 100, and 411 is about 400.
   100 + 400 = 500
   Actual Answer: 548
2. 226 is about 200, and 312 is about 300.
   200 + 300 = 500
   Actual Answer: 538
3. 538 is about 500, and 190 is about 200.
   500 + 200 = 700
   Actual Answer: 728
4. 315 is about 300, and 231 is about 200.
   300 + 200 = 500
   Actual Answer: 546
5. 722 is about 700, and 133 is about 100.
   700 + 100 = 800
   Actual Answer: 855
6. 302 is about 300, and 593 is about 600.
   300 + 600 = 900
   Actual Answer: 895
7. 781 is about 800, and 110 is about 100.
   800 + 100 = 900
   Actual Answer: 891
8. 613 is about 600, and 324 is about 300.
   600 + 300 = 900
   Actual Answer: 937
9. 596 is about 600, and 212 is about 200.
   600 + 200 = 800
   Actual Answer: 808
10. 479 is about 500, and 222 is about 200.
    500 + 200 = 700
    Actual Answer: 701

PART 3

1. 40 + 21 = 61
2. 12 + 31 = 43
3. 45 + 23 = 68
4. 60 + 16 = 76
5. 34 + 54 = 88

PART 4

1. **a.** The problem is asking: Did Hector and Ernesto win the game?
   b. Estimate: 4,396 is about 4,000, and 5,087 is about 5,000.
   4,000 + 5,000 = 9,000
   From the estimate, it does not look like they won the game.

c. Actual Answer: 4,396 + 5,087 = 9,483
This is less than the next closest team, who have 9,975 points. Therefore, the answer is that Hector and Ernesto did not win the game.

2. **a.** The problem is asking: How many students climbed the wall successfully during the past two years?
b. Estimate: 472 is about 500, and 429 is about 400.
500 + 400 = 900
From the estimate, it looks like about 900 students climbed the wall successfully during the past two years.
c. Actual Answer: 472 + 429 = 901
There were 901 students who successfully climbed the wall during the past two years.

Lesson 18—Constructing Graphs

See individual student work. Answers will vary according to the information collected by the class.

Be sure the answers include the following:

1. A correctly constructed graph that clearly compares the data collected.
2. Clearly identified titles and unit labels.
3. A paragraph that compares and contrasts the data—what's the same, what's the difference—using good math language.

Lesson 18—Homework

PART 1

1. 15
2. 150
3. 1,500
4. 15
5. 15
6. 150
7. 1,500
8. 13
9. 130
10. 1,300
11. 13
12. 13
13. 13
14. 130
15. 130
16. 1,300

PART 2

1. 200 + 20 + 8 + 40 + 1 =
200 + 0 + 20 + 40 + 8 + 1 =
200 + 60 + 9 = 269
2. 500 + 30 + 4 + 30 + 4 =
500 + 0 + 30 + 30 + 4 + 4 =
500 + 60 + 8 = 568
3. 300 + 90 + 0 + 100 + 0 + 3 =
300 + 100 + 90 + 0 + 0 + 3 =
400 + 90 + 3 = 493
4. 200 + 10 + 6 + 10 + 2 =
200 + 0 + 10 + 10 + 6 + 2 =
200 + 20 + 8 = 228
5. 400 + 0 + 7 + 90 + 2 =
400 + 0 + 0 + 90 + 7 + 2 =
400 + 90 + 9 = 499
6. 700 + 0 + 1 + 100 + 20 + 6 =
700 + 100 + 0 + 20 + 1 + 6 =
800 + 20 + 7 = 827
7. 100 + 0 + 2 + 90 + 0 =
100 + 0 + 0 + 90 + 2 + 0 =
100 + 90 + 2 = 192
8. 400 + 20 + 0 + 100 + 0 + 9 =
400 + 100 + 20 + 0 + 0 + 9 =
500 + 20 + 9 = 529
9. 600 + 80 + 0 + 200 + 0 + 7 =
600 + 200 + 80 + 0 + 0 + 7 =
800 + 80 + 7 = 887

PART 3

1. 20 + 40 + 5 + 4 = 60 + 9 = 69
25 + 44 = 69
2. 60 + 20 + 1 + 2 = 80 + 3 = 83
61 + 22 = 83
3. 80 + 10 + 4 + 5 = 90 + 9 = 99
84 + 15 = 99
4. 100 + 200 + 60 + 10 + 7 + 0 =
300 + 70 + 7 = 377
167 + 210 = 377
5. 600 + 100 + 0 + 40 + 5 + 2 =
700 + 40 + 7 = 747
605 + 142 = 747
6. 300 + 0 + 40 + 50 + 1 + 2 =
300 + 90 + 3 = 393
341 + 52 = 393

Lesson 19—Making Pictographs

1. Sample Answer:

Money Made on One Movie

Miramax	$$$	$$$	$$$	$$$	$			
Fox	$$$	$$$	$$$	$$				
Paramount	$$$	$$						
Disney	$$$	$$$	$$					
Dreamworks	$$$	$$$	$$$	$$$	$$$	$$		
Warner Bros.	$$$	$$$	$$$	$$$	$$$	$$$	$$$	$

KEY: $$$ = $5 million

2. $$$ is the symbol used to show money. It's worth $5 million.
3. Sample problems and answers based on the data:
   a. What movie studio earned the most money on one movie? What was the amount?
      Answer: Warner Brothers, $36 million
   b. What movie studio earned the least money on one movie? What was the amount?
      Answer: Paramount, $8 million

Lesson 19—Homework

PART 1

1. 2
2. 80
3. 200
4. 7
5. 30
6. 300
7. 50
8. 300
9. 2

PART 2

1. 40 + 8 + 30 + 7 ⟶ 48 + 37
2. 100 + 30 + 5 + 400 + 20 + 7 ⟶
   135 + 427
3. 60 + 7 + 20 + 3 ⟶ 67 + 23
4. 300 + 1 + 400 + 90 ⟶ 301 + 490

PART 3

1. 10 + 2 + 50 + 6
   10 + 50 + 2 + 6
   60 + 8
   68
2. 100 + 50 + 0 + 20 + 9
   100 + 50 + 20 + 0 + 9
   100 + 70 + 9
   179
3. 200 + 0 + 7 + 500 + 90 + 1
   200 + 500 + 0 + 90 + 7 + 1
   700 + 90 + 8
   798
4. 40 + 5 + 30 + 4
   40 + 30 + 5 + 4
   70 + 9
   79
5. 200 + 10 + 4 + 80 + 3
   200 + 10 + 80 + 4 + 3
   200 + 90 + 7
   297
6. 500 + 10 + 2 + 400 + 80 + 7
   500 + 400 + 10 + 80 + 2 + 7
   900 + 90 + 9
   999
7. 20 + 4 + 10 + 3
   20 + 10 + 4 + 3
   30 + 7
   37
8. 100 + 20 + 3 + 300 + 20 + 4
   100 + 300 + 20 + 20 + 3 + 4
   400 + 40 + 7
   447
9. 300 + 30 + 3 + 500 + 40 + 5
   300 + 500 + 30 + 40 + 3 + 5
   800 + 70 + 8
   878

PART 4

1. There were 100 CDs sold in January.
2. There were 400 CDs sold in April.
3. There were 300 more CDs sold in April.
   100 + 300 = 400
4. The trend is that sales are increasing by 100 each month.
5. From the trend you have seen, a reasonable guess would be $500.

Lesson 20—Homework

PART 1

1. 50
2. 2
3. 2,000
4. 700

PART 2

1. The problem is asking: Was Toby riding fast enough to get into the Guinness Book of Records?
2. The problem is asking: How much does the world's population increase each month?
3. The problem is asking: How long will it take a window-washing crew to wash all the windows on 1 side of the skyscraper?

PART 3

1. 698 ⟶ 600 + 90 + 8
2. 1,025 ⟶ 1,000 + 0 + 20 + 5
3. 506 ⟶ 500 + 0 + 6
4. 430 ⟶ 400 + 30 + 0
5. 413 ⟶ 400 + 10 + 3

PART 4

1.

10	
90	7
+ 20	4
120	1

121

2.

10	
80	9
+ 10	5
100	4

104

3.

100	10	
100	40	7
+	70	6
200	20	3

223

4.

100	10	
200	60	2
+ 100	30	8
400	0	0

400

PART 5 (A range of acceptable answers is given.)

1. 1–3
2. 640–660

PART 6

1. $\underline{30 + 9} + \underline{40 + 0}$
   $\underline{30 + 40} + \underline{9 + 0}$
   $70 + 9$
   79
2. $\underline{40 + 2} + \underline{10 + 6}$
   $\underline{40 + 10} + \underline{2 + 6}$
   $50 + 8$
   58
3. $\underline{60 + 1} + \underline{10 + 5}$
   $\underline{60 + 10} + \underline{1 + 5}$
   $70 + 6$
   76
4. $\underline{20 + 4} + \underline{50 + 3}$
   $\underline{20 + 50} + \underline{4 + 3}$
   $70 + 7$
   77

PART 7

1. 38 is about 40, and 51 is about 50.
   $40 + 50 = 90$
   Actual Answer: 89
2. 49 is about 50, and 17 is about 20.
   $50 + 20 = 70$
   Actual Answer: 66
3. 187 is about 200, and 112 is about 100.
   $200 + 100 = 300$
   Actual Answer: 299

STUDENT TEXTBOOK ANSWER KEY

Subtraction

UNIT 2

Lesson 21—Homework

PART 1

1.

80	9
− 20	1
60	8

68

2.

20	9
− 10	1
10	8

18

3.

90	7
− 60	4
30	3

33

4.

70	8
− 10	5
60	3

63

5.

40	7
− 20	0
20	7

27

6.

50	3
− 10	2
40	1

41

7.

60	7
− 30	6
30	1

31

8.

70	7
− 40	4
30	3

33

9.

30	9
− 20	8
10	1

11

PART 2

1. $3 + 9 = 12$
$9 + 3 = 12$
$12 - 3 = 9$
$12 - 9 = 3$

2. $6 + 7 = 13$
$7 + 6 = 13$
$13 - 6 = 7$
$13 - 7 = 6$

3. $8 + 6 = 14$
$6 + 8 = 14$
$14 - 8 = 6$
$14 - 6 = 8$

4. $9 + 8 = 17$
$8 + 9 = 17$
$17 - 9 = 8$
$17 - 8 = 9$

PART 3 (Items in bold type are given in the text.)

Ones	Tens	Hundreds
17 − 8 = 9	**170 − 80 = 90**	**1,700 − 800 = 900**
12 − 6 = 6	**120 − 60 = 60**	1,200 − 600 = 600
13 − 5 = 8	130 − 50 = 80	1,300 − 500 = 800
11 − 4 = 7	110 − 40 = 70	**1,100 − 400 = 700**
15 − 9 = 6	150 − 90 = 60	**1,500 − 900 = 600**
14 − 6 = 8	140 − 60 = 80	1,400 − 600 = 800
16 − 9 = 7	**160 − 90 = 70**	1,600 − 900 = 700

Lesson 22—Reading Information From a Bar Graph

1. $300 - 200 = 100$
2. $200 - 100 = 100$
3. $100 + 200 = 300$
4. $300 + 400 = 700$
5. $700 - 300 = 400$

Lesson 22—Reading Data From a Table

Sample Responses (Check individual student work.)

1. How much greater were sales in April than in January?
   Answer: $400 - 100 = 300$. Sales were 300 more in April.
2. What were the total sales for the four-month period?
   Answer: $100 + 200 + 300 + 400 = 1{,}000$ total sales

Lesson 22—Homework

PART 1

1.

90	8
− 60	4
30	4

34

2.

70	6
− 50	3
20	3

23

3.

70	5
− 50	3
20	2

22

4.

50	3
− 30	1
20	2

22

5.

30	6
− 20	5
10	1

11

6.

90	70
− 60	5
30	2

32

7.

20	4
− 10	2
10	2

12

8.

40	9
− 20	8
20	1

21

PART 2

1. 30 + 90 = 120
   90 + 30 = 120
   120 − 30 = 90
   120 − 90 = 30
2. 80 + 60 = 140
   60 + 80 = 140
   140 − 80 = 60
   140 − 60 = 80
3. 40 + 70 = 110
   70 + 40 = 110
   110 − 70 = 40
   110 − 40 = 70
4. 600 + 700 = 1,300
   700 + 600 = 1,300
   1,300 − 600 = 700
   1,300 − 700 = 600
5. 900 + 800 = 1,700
   800 + 900 = 1,700
   1,700 − 900 = 800
   1,700 − 800 = 900
6. 700 + 900 = 1,600
   900 + 700 = 1,600
   1,600 − 900 = 700
   1,600 − 700 = 900

PART 3

1. 8
2. 9
3. 4
4. 80
5. 90
6. 40

PART 4

1. May—100, June—200, July—300, August—400
2. 200 − 100 = 100
3. 400 − 300 = 100
4. 100 + 200 = 300; 300 + 400 = 700; 700 − 300 = 400

Lesson 23—Word Problems

1. Which group of cities drew the most attendance? Answer: The first group. What was the difference?
   Answer: Estimate 16,000 − 13,000 = 3,000
   Actual Answer: 15,612 − 12,999 = 2,613
2. What was the attendance figure for the last concert?
   Answer: Estimate 11,000 − 10,000 = 1,000
   Actual Answer: 11,223 − 9,899 = 1,324 attendance at last concert
3. What was the total difference in attendance for the two bands for the summer tour?
   Blockheads' total: 15,612 + 12,999 = 28,611
   Toasters' total: 11,223
   Difference: 28,611 − 11,223 = 17,388
4. How many more fans did the Jaylands draw than the Toasters? How many more than the Blockheads?
   Jaylands' total: 28,766
   Toasters' total: 11,223
   Difference: 28,766 − 11,223 = 17,543
   Jaylands' total: 28,766
   Blockheads' total: 28,611
   Difference: 28,766 − 28,611 = 155

Lesson 23—Homework

PART 1

1.

60	9
− 20	1
40	8

48

2.

400	20	9
− 300	10	1
100	10	8

118

3.

600	90	7
− 300	60	4
300	30	3

333

4.

70	8
− 10	5
60	3

63

5.

100	40	7
− 100	20	0
	20	7

27

6.

700	50	3
− 500	10	2
200	40	1

241

7.

50	7
− 30	6
20	1

21

8.

800	70	7
− 400	40	4
400	30	3

433

9.

200	30	9
− 100	20	8
100	10	1

111

PART 2 (Items in bold type are given in the text.)

Ones	Tens	Hundreds
13 − 8 = 5	**130 − 80 = 50**	**1,300 − 800 = 500**
11 − 6 = 5	**110 − 60 = 50**	1,100 − 600 = 500
12 − 5 = 7	120 − 50 = 70	1,200 − 500 = 700
14 − 7 = 7	140 − 70 = 70	**1,400 − 700 = 700**
17 − 9 = 8	170 − 90 = 80	**1,700 − 900 = 800**
15 − 6 = 9	150 − 60 = 90	1,500 − 600 = 900
18 − 9 = 9	**180 − 90 = 90**	1,800 − 900 = 900

PART 3

1. 30 + 80 = 110
   80 + 30 = 110
   110 − 30 = 80
   110 − 80 = 30
2. 800 + 300 = 1,100
   300 + 800 = 1,100
   1,100 − 300 = 800
   1,100 − 800 = 300
3. 70 + 90 = 160
   90 + 70 = 160
   160 − 90 = 70
   160 − 70 = 90
4. 900 + 700 = 1,600
   700 + 900 = 1,600
   1,600 − 700 = 900
   1,600 − 900 = 700

PART 4 (See individual student work to check the correct use of number lines.)

1. Estimate: 100 − 60 = 40
   Exact Answer: 41
2. Estimate: 600 − 300 = 300
   Exact Answer: 336
3. Estimate: 50 − 30 = 20
   Exact Answer: 18
4. Estimate: 900 − 100 = 800
   Exact Answer: 818
5. Estimate: 50 − 10 = 40
   Exact Answer: 39
6. Estimate: 600 − 300 = 300
   Exact Answer: 263

PART 5

1. Estimate: 14,000 − 7,000 = 7,000
   Exact Answer: 13,513 − 6,789 = 6,724
2. What band sold the most CDs? Livin'
   How many more CDs than the Sensors?
   Estimate: 1,700 − 900 = 800
   Exact Answer: 1,712 − 898 = 814
   How many more CDs than Uriah?
   Estimate: 1,700 − 500 = 1,200
   Exact Answer: 1,712 − 499 = 1,213
   Did Livin' sell more CDs than both of the other bands combined? Yes
   Estimates:
   Sensors: 898 ⟶ 900
   Uriah: 499 ⟶ + 500
   1,400
   Livin': 1,712 ⟶ 1,700
   Estimate: 1,700
   − 1,400
   300
   Exact Answers:
   Combined = 898 + 499 = 1,397
   How many more? 1,712 − 1,397 = 315 more

Lesson 24—Reading Data in Tables and Making Graphs

Sample Response (See individual student work.)

1. A bar graph can be made based on the actors and their asking salaries. It would include the information in the "Name" column of the table and the information in the "Asking Salary (in Millions)" column.
2. Here is a sample bar graph for that information:

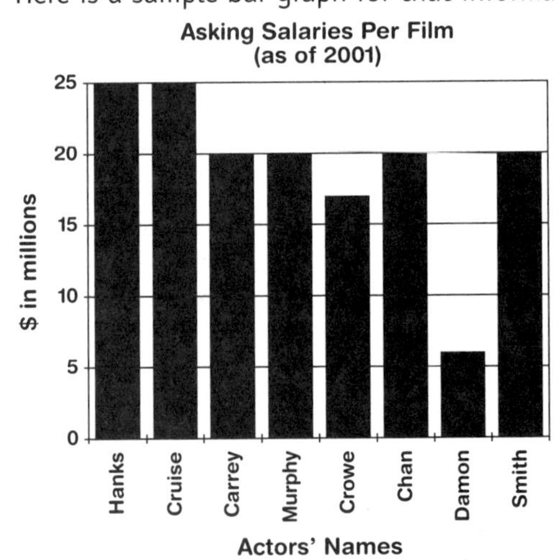

3. Some sample problems based on the information in the bar graph include the following:
   a. Who has the highest asking salary?
      Answer: Hanks and Cruise
      How much is that salary? $25,000,000
   b. Who has the lowest asking salary?
      Answer: Damon
      How much is that salary? $6,000,000
   c. What is the difference between the asking salaries of the highest paid actors and the lowest paid actors?
      $25,000,000 − $6,000,000 = $19,000,000

Lesson 24—Homework
PART 1

1.

	60	9
−	20	1
	40	8

48

2.

	400	80	9
−		10	1
	400	70	8

478

3.

	600	20	4
−	300	10	2
	300	10	2

312

4.

	300	20	8
−		10	5
	300	10	3

313

5.

	900	40	7
−		30	5
	900	10	2

912

6.

	800	0	8
−	200	0	2
	600	0	6

606

7.

	200	50	6
−		30	6
	200	20	0

220

8.

	200	70	7
−		40	2
	200	30	5

235

9.

	700	80	9
−	500	40	8
	200	40	1

241

PART 2

1. 900
2. 50
3. 7
4. 600
5. 20
6. 200
7. 90
8. 70
9. 60
10. 500

PART 3

1. 40 + 80 = 120
   80 + 40 = 120
   120 − 40 = 80
   120 − 80 = 40
2. 400 + 800 = 1,200
   800 + 400 = 1,200
   1,200 − 400 = 800
   1,200 − 800 = 400
3. 90 + 40 = 130
   40 + 90 = 130
   130 − 90 = 40
   130 − 40 = 90
4. 900 + 400 = 1,300
   400 + 900 = 1,300
   1,300 − 900 = 400
   1,300 − 400 = 900

PART 4

Check individual student work for the correct location of the numbers on number lines with different scales of 10, 25, and 100.

Lesson 25—Advertising on Broadcast TV, Cable TV, and the Internet

Sample Answers

1. about 17 billion + about 10 billion + about 4 billion = about $31 billion
2. about 33 billion − about 4 billion = about $29 billion
3. Broadcast TV: about 19 billion − about 17 billion − about $2 billion change (compared to about $14 billion difference in cable and about $29 billion in the Internet)
4. Sample question ⟶ About how many billion dollars predicted to be spent in advertising in 2005 on Broadcast TV, Cable TV and Internet combined?
   Answer: 19 + 24 + 33 = about 20 + 20 + 30 = about $70 billion

Lesson 25—Homework
PART 1
1. Estimate: 60 − 30 = 30
   Exact Answer: 33
2. Estimate: 80 − 20 = 60
   Exact Answer: 63
3. Estimate: 30 − 20 = 10
   Exact Answer: 16
4. Estimate: 40 − 20 = 20
   Exact Answer: 25
5. Estimate: 50 − 40 = 10
   Exact Answer: 14
6. Estimate: 80 − 50 = 30
   Exact Answer: 29

PART 2
1. Estimate: 225 − 125 = 100
   Exact Answer: 97
2. Estimate: 475 − 325 = 150
   Exact Answer: 155
3. Estimate: 325 − 100 = 225
   Exact Answer: 225
4. Estimate: 950 − 25 = 925
   Exact Answer: 919
5. Estimate: 875 − 350 = 525
   Exact Answer: 520
6. Estimate: 675 − 400 = 275
   Exact Answer: 276

Lesson 26—The Blockheads Concert Tour: First Stop, Pennsylvania
Sample Answers (Check individual student work.)
1. **a.** 75 + 150 + 125 + 50 = 400 ⟶ about 400 miles going through Williamsport
   b. 200 + 100 = 300 ⟶ about 300 miles going through Harrisburg
2. 400 − 300 = 100 ⟶ about 100 miles difference between the 2 ways
3. Calculator answers:
   Going through Williamsport: Actual distance = 415 miles
   My estimate of 400 was very close ⟶ only 5 miles off
   Going through Harrisburg: Actual distance = 311 miles
   My estimate of 300 was very close ⟶ only 11 miles off
   Actual difference ⟶ 415 − 311 = 104 miles
   My estimate of 100 was only 4 miles off. That's very close!

   I used quarter rounding for this problem. I think that was a good method because all of my estimates were very close.

Lesson 26—Homework
PART 1
1.	75	75	750
2.	125	125	125
3.	50	50	500
4.	50	50	500

PART 2
1. Front-End: 500 − 300 = 200
   Estimate: 200
   Actual Answer: 202
2. Quarter: 850 − 125 = 725
   Estimate: 725
   Actual Answer: 730
3. Quarter: 7,500 − 5,000 = 2,500
   Estimate: 2,500
   Actual Answer: 2,417
4. Front-End: 2,000 − 1,000 = 1,000
   Estimate: 1,000
   Actual Answer: 805
5. Front-End: 1,600 − 800 = 800
   Estimate: 800
   Actual Answer: 800
6. Quarter: 7,500 − 2,500 = 5,000
   Estimate: 5,000
   Actual Answer: 5,000

PART 3
1. Estimate: 60 − 40 = 20
   Actual Answer: 23
2. Estimate: 350 − 50 = 300
   Actual Answer: 309
3. Estimate: 60 − 30 = 30
   Actual Answer: 38
4. Estimate: 750 − 525 = 225
   Actual Answer: 218
5. Estimate: 850 − 50 = 800
   Actual Answer: 793
6. Estimate: 900 − 650 = 250
   Actual Answer: 281

PART 4
1. Which band received the most votes?
   The Blockheads
   Estimate: 475 − 450 = 25
   What is the difference between first and second place? 477 − 457 = 20 votes difference

2. Which bands would Kazoodle be interested in? The Road Rats and the Toasters
   What band is the closest to 800 that did not make the cut? The Blockheads
   How many more CDs? 200 more
   (600 + 200 = 800)

2. Did Jon get a bonus during any of the months listed?
   Yes, he got a bonus in January and March.
   How many more sales did he need in February to get to 100?
   He needed 12 more (100 − 88 = 12).

Lesson 27—How Much Did They Make in Pennsylvania?

1. About how much did the Blockheads spend in Pennsylvania?
   500 + 300 + 1,000 + 500 + 100 = 2,400
   ⟶ About $2,400
2. About how much were the Blockheads paid?
   3,000 + 4,000 = 7,000 ⟶ About $7,000
3. About how much were their profits?
   $7,000 − $2,400 = $4,600 ⟶ About $4,600
4. Exact calculations:
   Expenses: 478 + 294 + 986 + 461 + 94 = 2,313
   Differs from estimate by $87
   Payments: 3,250 + 3,600 = 6,850
   Differs from estimate by $150
   Profits: $6,850 − $2,313 = $4,537
   Differs from estimate by $63
   BONUS: $4,537 divided by 5 band members is $907.40 each.

Lesson 27—Homework

PART 1

1. 39
2. 491
3. 37
4. 362
5. 208
6. 271

PART 2

1. The circled part shows the ones place with a borrowed ten. You add the ten to the digit in the ones place ⟶ 10 + 4 = 14.
2. The circled part shows the adjustment made in the tens place when you borrow. You cross out the digit and subtract one. Here, you crossed out the 5 and made it a 4.

PART 3

1. What theme received the most votes?
   Medieval Times
   Estimate: 470 − 410 = 60
   What was the difference between first and second place? 474 − 413 = 61
   A difference of 61.

Lesson 28—Homework

PART 1

See individual student work for expanded form of each problem. Answers only are given here.

1. 35
2. 135
3. 32
4. 342
5. 516
6. 335
7. 411
8. 555

PART 2 (Items in bold type are given in the text.)

Ones	Tens	Hundreds
12 − 8 = 4	**120 − 80 = 40**	**1,200 − 800 = 400**
13 − 6 = 7	**130 − 60 = 70**	1,300 − 600 = 700
14 − 5 = 9	140 − 50 = 90	1,400 − 500 = 900
15 − 7 = 8	150 − 70 = 80	**1,500 − 700 = 800**
16 − 9 = 7	160 − 90 = 70	**1,600 − 900 = 700**
17 − 8 = 9	170 − 80 = 90	1,700 − 800 = 900
18 − 9 = 9	**180 − 90 = 90**	1,800 − 900 = 900

PART 3

1. 9 + 6 = 15
   6 + 9 = 15
   15 − 6 = 9
   15 − 9 = 6
2. 4 + 7 = 11
   7 + 4 = 11
   11 − 7 = 4
   11 − 4 = 7
3. 90 + 60 = 150
   60 + 90 = 150
   150 − 90 = 60
   150 − 60 = 90
4. 400 + 700 = 1,100
   700 + 400 = 1,100
   1,100 − 400 = 700
   1,100 − 700 = 400

PART 4

1. Estimate: 50 − 20 = 30
   Actual Answer: 29
2. Estimate: 200 − 100 = 100
   Actual Answer: 115
3. Estimate: 70 − 50 = 20
   Actual Answer: 24
4. Estimate: 700 − 400 = 300
   Actual Answer : 292
5. Estimate: 550 − 50 = 500
   Actual Answer: 486
6. Estimate: 450 − 125 = 325
   Actual Answer: 329

7. Estimate: 625 − 225 = 400
   Actual Answer: 403
8. Estimate: 900 − 400 = 500
   Actual Answer: 485

Lesson 29—The Blockheads Concert Tour: Florida

Decide which car they should rent. Student answers should include the following information:

- Discussion about the type of car.
  Because there are 5 members in the band, they can choose from only the midsize and the luxury car.
- Discussion about miles per gallon.
  Students should make observations about which of the two cars gets the best gas mileage: the midsize, 22 mpg; the luxury, 18 mpg.
- Discussion about one-way daily rate.
  Students should compare the one-way daily rates for the two types of cars to see which would be the best value for their money: midsize, $97 per day; the luxury, $109 per day.

Questions about the drive from Orlando to Miami.

1. Through Sarasota: 80 + 60 + 200 = about 340 miles
2. Through West Palm Beach: 170 + 70 = about 240 miles
3. About 110 miles difference between the two routes.
4. If they go through Sarasota, they'll have to pay for 150 miles at 10 cents per mile. If they go through West Palm Beach, they'll have to pay for 40 miles at 10 cents per mile.
5. Exact calculations:
   Exact mileage through Sarasota ⟶ 85 + 61 + 231 = 377 (estimate was 27 miles off)
   Exact mileage through West Palm Beach ⟶ 171 + 73 = 244 (estimate was 4 miles off)
   Exact difference in mileage between two routes ⟶ 377 − 244 = 133 (estimate was 23 miles off)
   BONUS: If they go through Sarasota and they rent the midsize car:
   They pay $97 per day for the midsize car.
   The trip is 377 miles, so they have to pay 10 cents per mile for 177 miles.
   10 cents x 177 = $17.70.
   Add $17.70 to $97 and you get $114.70.

Lesson 29—Homework

PART 1

1.	9	90	900
2.	9	900	90
3.	25	25	250
4.	7	70	700

PART 2

1. The correct answer is 202. The error is in the ones place. The student switched the digits around. Instead of 1 − 9, which involves borrowing from the tens place, the student subtracted 9 − 1 = 8.
2. The correct answer is 296. This is 100 less than the answer the student got. We see that the student forgot to adjust the hundreds place after borrowing 100.
3. The correct answer is 121. The error is in the tens place. The student switched the digits around. Instead of 0 − 8, which involves borrowing from the hundreds place, the student subtracted 8 − 0 = 8.

PART 3

1. What was the attendance at the third show?
   First show: 1,058 (estimate 1,000)
   Second show: 959 (estimate 1,000)
   Total of all shows: 3,128 (estimate 3,000)
   Add together first and second:
   Estimate 1,000 + 1,000 = 2,000
   (Calculator Answer: 1,058 + 959 = 2,017)
   Subtract from total:
   Estimate 3,000 − 2,000 = 1,000
   (Calculator Answer: 3,128 − 2,017 = 1,111)
   Attendance at the third show was 1,111.
2. What were their sales for February?
   Total Sales for Jan.–March: 1,750 (estimate 1,800)
   Jan. Sales: 575 (estimate 600)
   March Sales: 625 (estimate 600)
   Add together Jan. and March:
   Estimate 600 + 600 = 1,200
   (Calculator Answer: 575 + 625 = 1,200)
   Subtract from total:
   Estimate: 1,700 − 1,200 = 500
   (Calculator Answer: 1,750 − 1,200 = 550)
   Feb. Sales were 550.

Lesson 30—Class Activity: Checking the Bill

1. Rounding each of the numbers and finding the sum:
   100 + 10 + 10 + 100 + 10 + 10 + 100 + 10 + 10 + 100 + 10 + 10 = 480
   The estimate of 480 makes the total of $648.30 seem high.
2. Using a calculator to find the exact total: $538.80
   Does it match the bill? No, it does not match.
3. There seems to be a mistake. The difference is $109.50. That's the cost of a room with 2 queen beds. It looks like the clerk added that charge an extra time.

Lesson 30—Homework

PART 1

1. 7	70	700
2. 7	70	700
3. 8	80	800
4. 8	80	800

PART 2

Problem 2 has the error. The answer should be 103. It looks like the person switched the digits in the ones place. This person should have subtracted 1 − 8, which would involve borrowing from the tens place. Instead, he or she just subtracted 8 − 1 = 7.

PART 3

1. Estimate: 350 − 225 = 125
   Actual Answer: 122
2. Estimate: 900 − 300 = 600
   Actual Answer: 578
3. Estimate: 400 − 200 = 200
   Actual Answer: 173
4. Estimate: 675 − 425 = 250
   Actual Answer: 254
5. Estimate: 950 − 725 = 225
   Actual Answer: 227
6. Estimate: 775 − 225 = 550
   Actual Answer: 552

PART 4

1. (a) traditional subtraction: 7,698
2. (b) estimation and calculator:
   5,000 − 1,000 = 4,000 ⟶ 4,189
3. (a) traditional subtraction: 5,487

Lesson 31—Figuring Out How Much You Made

Problem 1: There is no problem here.
Problem 2: How much should the Blockheads be paid?
Estimate: 4,000 − 3,000 = 1,000
Exact Answer: 4,215 − 2,843 = $1,372
Problem 3: How much was their profit?
Estimate: 500 + 250 = 750
Exact Answer: 510 + 240 = $750 total costs
Estimate: 1,200 − 750 = 450
Exact Answer: 1,219 − 750 = $469 profit

Lesson 31—Homework

PART 1

Problem 1 has the error. Instead of regrouping, the person simply switched the numbers in the ones place. He or she subtracted 7 − 6 = 1 instead of 6 − 7, which requires borrowing from the tens place.

PART 2

1. 7 + 9 = 16
   9 + 7 = 16
   16 − 7 = 9
   16 − 9 = 7
2. 25 + 75 = 100
   75 + 25 = 100
   100 − 25 = 75
   100 − 75 = 25
3. 90 + 70 = 160
   70 + 90 = 160
   160 − 70 = 90
   160 − 90 = 70
4. 75 + 25 = 100
   25 + 75 = 100
   100 − 25 = 75
   100 − 75 = 25
5. 900 + 700 = 1,600
   700 + 900 = 1,600
   1,600 − 700 = 900
   1,600 − 900 = 700
6. 500 + 500 = 1,000
   500 + 500 = 1,000
   1,000 − 500 = 500
   1,000 − 500 = 500

PART 3

1. (a) traditional subtraction: 467
2. (b) estimation and calculator:
   900 − 200 = 700 ⟶ 699
3. (b) estimation and calculator:
   6,000 − 1,000 = 5,000 ⟶ 5,085
4. (b) estimation and calculator:
   5,000 − 1,000 = 4,000 ⟶ 4,189
5. (b) estimation and calculator:
   225 − 200 = 25 ⟶ 25
6. (a) traditional subtraction: 5,487

Lesson 32—On to Texas

1. The Blockheads had budgeted $1,000 for transportation in Texas.
   a. How much did they spend on travel expenses from Houston to El Paso?
   They spent about $1,000. 600 + 200 + 100 + 100 = 1,000.
   Actual cost: 645.79 + 230 + 61.31 + 105 = 937.10 + 105 = $1,042.10
   b. Were they over or under budget? By how much?
   They were over budget by $42.10.
2. They drove about 1,100 miles (200 + 100 + 200 + 600).
   Actual mileage is 1,112 miles.
3. Cost = $1,042.10
   Mileage = 1,112 miles
4. It's 554 miles from San Antonio to El Paso.
   a. How far? ⟶ 200 + 600 = 800
   b. How much shorter? ⟶ 1,100 − 800 = 300
   c. Actual distance traveled: 751 miles
   Actual distance shorter: 1,112 − 751 = 361

Lesson 32—Homework

PART 1

1. Estimate: 400 − 200 = 200
   Exact Answer: 188
2. Estimate: 700 − 400 = 300
   Exact Answer: 289
3. Estimate: 7,000 − 2,000 = 5,000
   Exact Answer: 4,888
4. Estimate: 575 − 400 = 175
   Exact Answer: 176
5. Estimate: 900 − 200 = 700
   Exact Answer: 709
6. Estimate: 5,000 − 4,000 = 1,000
   Exact Answer: 1,008

PART 2

1. 9
2. 9,000
3. 25
4. 250
5. 500
6. 2
7. 2,000
8. 250
9. 7
10. 700

PART 3

1. (a) traditional subtraction: 4,455
2. (b) estimation and calculator:
   8,000 − 2,000 = 6,000 ⟶ 6,089
3. (a) traditional subtraction: 251

Lesson 33—The Bottom Line: Did We Make Any Money?

Students are to make up three problems about the data and provide answers to the problems.

Answers to their problems may include some of the following figures:

1. Estimate of Expenses: $13,000
   Actual Expenses: $13,034
2. Estimate of Payments: $30,000
   Actual Payments: $30,081
3. Profits (Payments − Expenses): $17,047
   BONUS: Students are to make a bar graph illustrating some of their information.
   See individual student work. Be sure the graph includes titles, labels, and accurate data.

Lesson 33—Homework

PART 1

1. 1
2. 5
3. 10

PART 2

1. 15
   150
   1,500
2. 1
   1
   1
3. 14
   140
   1,400
4. 9
   99
   999

PART 3

1. (a) traditional subtraction: 111
2. (c) mental math: 5
3. (b) estimation and calculator:
   2,100 − 2,000 = 100 ⟶ 136
4. (c) mental math: 26 (Think 1 more makes 2,775 and 25 more makes 3,000.)
5. (b) estimation and calculator:
   675 − 450 = 225 ⟶ 225
6. (a) traditional subtraction: 411

Lesson 34—How Much Does a Movie Make? Gross Profits for *Shrek*

1. Students are to write three to four good questions about the information given. The following is some information that may be included in the student answers:
   Totals for each week:
   Week 1 ⟶ $60 million
   Week 2 ⟶ $60 million

Week 3 ⟶ $40 million
Week 4 ⟶ $30 million
Week 5 ⟶ $20 million
Week 6 ⟶ $20 million
Week 7 ⟶ $10 million
Week 8 ⟶ $10 million
Week 9 ⟶ $10 million
Week 10 ⟶ $10 million

2. With regard to trends, students should note that the profits go up after the first week, then gradually go down each week after that.

Lesson 34—Homework
PART 1
1. 2
2. 203
3. 1
4. 6,001
5. 1
6. 999

PART 2
1. 20,000 − 19,499 = 501 more sold this year.
2. 19,712 − 18,510 = 1,202 more sold this year.
   Two-year total for Kazoom:
   19,499 + 20,000 = 39,499 (estimate 40,000)
   Two-year total for Kazoodle:
   18,510 + 19,712 = 38,222 (estimate 40,000)
   Which company sold the most? Kazoom
   By how much? 39,499 − 38,222 = 1,277 more

Lesson 35—Homework
PART 1
Check individual student work for expanded form of the problems. The answers to the problems are as follows.
1. 76
2. 111
3. 231
4. 611

PART 2
Check individual student work for expanded form of the problems. The answers to the problems are as follows.
1. 63
2. 503
3. 23
4. 506
5. 26
6. 241

PART 3
1. 3
2. 8
3. 5
4. 30
5. 80
6. 500
7. 300
8. 800
9. 50

PART 4
1. Front-End rounding
   Estimate: 90 − 60 = 30
   Exact Answer: 32
2. Quarter rounding
   Estimate: 850 − 225 = 625
   Exact Answer: 620
3. Front-End rounding
   Estimate: 600 − 200 = 400
   Exact Answer: 398

PART 5
1. The correct answer is 29. The error is in the ones place. The person solving the problem was supposed to subtract 5 − 6, which requires borrowing from the tens place. Instead, he or she subtracted 6 − 5 = 1.
2. The correct answer is 24. The person solving the problem forgot to adjust the tens place down by 1 after borrowing 10 and adding it in the ones place.
3. The correct answer is 92. The error is in the tens place. The person solving the problem should have subtracted 8 − 9, which requires borrowing from the hundreds place. Instead, he or she subtracted 9 − 8 = 1.

PART 6
1. 1
2. 6,014
3. 3,124
4. 98
5. 1
6. 999,999

STUDENT TEXTBOOK ANSWER KEY

Multiplication

UNIT 3

Lesson 36—Estimating Distance

Problem 1: Students should use the skateboard as the unit of measurement.

1. It's about 2 or 3 skateboards from A to B. Assuming that a skateboard is about 2 or 3 feet, the estimate should be somewhere between 4 and 9 feet ($2 \times 2 = 4$ feet is the low end; $3 \times 3 = 9$ feet is the high end).
2. Look at individual student responses. Look for discussions similar to that given here.

Problem 2: Students may use a couple different strategies.

Strategy 1: Students may use the diver and the stairs as the unit of measurement. The diver's length is about 6 stairs long. We can estimate the stairs as being about 12 inches apart.

Strategy 2: The diver's length is about 6 stairs long. That's about $6 \times 12 = 72$ inches. It takes about 2 diver lengths to get from C to D. $72 \times 2 = 144$ inches, or about 12 feet.

Strategy 3: The diver's length is about 6 feet tall. It takes about 2 diver lengths from C to D. $2 \times 6 = 12$ feet.

Final Comment: Look at individual student responses. Look for discussions similar to that given here.

Lesson 36—Homework

PART 1

1. 16
2. 15
3. 5
4. 160
5. 1,500
6. 500
7. 8
8. 6
9. 13
10. 800
11. 60
12. 130

PART 2

1.	72	720	7,200
2.	56	560	5,600
3.	54	540	5,400
4.	32	320	3,200
5.	45	450	4,500
6.	24	240	2,400

PART 3

1. $50 \times 4 = 200$
2. $800 \times 6 = 4{,}800$

Lesson 37—More Distances

Problem 1: Use the distance from the head to the foot of the climber as the unit of measurement, and go diagonally from A to B.

1. It's about 2 of the climber between A and B. If he's about 5 – 6 feet, that's about 10 – 12 feet between A and B.
2. Look at individual student responses. Look for discussions similar to that given here.

Lesson 37—Word Problems

1. Estimate: $3 \times 80 = \$240$
   Actual Answer: $234
2. Estimate: $200 + 200 + 300 = 700$ ft.
   Actual Answer: 709 ft.
3. Estimate: $400 \times 5 = 2{,}000$ cars
   Actual Answer: 2,000 cars

Lesson 37—Homework

PART 1

1. 72
2. 56
3. 24
4. 72
5. 20
6. 49
7. 42
8. 42
9. 36
10. 30
11. 81
12. 64

PART 2

1. 63 630 6,300
2. 48 480 4,800
3. 45 450 4,500
4. 42 420 4,200

PART 3

1. 7 × 10
2. 8 × 10
3. 9 × 10
4. 50 × 10
5. 40 × 10
6. 70 × 10
7. 800 × 10
8. 900 × 10
9. 19 × 10
10. 14 × 10

Lesson 38—Making a Measuring Device

The number of steps will vary depending on the length of the room.

See individual student work. Look for a discussion like this: "I would use my feet and step one foot after the next until I reached the end of the room. I would count the steps as I go."

Lesson 38—Homework

PART 1

1. 72
2. 24
3. 240
4. 720
5. 240
6. 2,400
7. 42
8. 4,200
9. 27
10. 4,200
11. 24
12. 270

PART 2 (Items in bold type are given in the text.)

The Starting Number	? × 10	? × 100	? × 1,000
1,000	**100 × 10**	10 × 100	**1 × 1,000**
2,000	200 × 10	**20 × 100**	2 × 1,000
3,000	**300 × 10**	30 × 100	**3 × 1,000**
6,000	**600 × 10**	**60 × 100**	6 × 1,000
5,000	500 × 10	**50 × 100**	5 × 1,000
9,000	900 × 10	90 × 100	9 × 1,000
4,000	400 × 10	40 × 100	4 × 1,000
7,000	700 × 10	**70 × 100**	**7 × 1,000**

PART 3

1. 2,000
2. 200
3. 10
4. 8,000
5. 8,000
6. 2,000
7. 1,000
8. 6
9. 8,000
10. 60

Lesson 39—Making a Ruler of 1s and 10s

See individual student work.

Lesson 39—Homework

PART 1

1.

	30	2
+	20	7
	50	9

59

2.

	40	3
+	10	5
	50	8

58

3.

	100	20	8
+		40	1
	100	60	9

169

4.

	200	60	5
+		30	1
	200	90	6

296

PART 2

1.

	30	5
−	20	1
	10	4

14

2.

	40	3
−	10	2
	30	1

31

3.

	500	80	7
−		70	6
	500	10	1

511

4.

	600	20	4
−	200	10	3
	400	10	1

411

PART 3

1.

	90	8
×		4
		32
+		360
		392

2.

	70	6
×		6
		36
+		420
		456

3.

	70	5
×		3
		15
+		210
		225

4.

	50	3
×		7
		21
+		350
		371

5.

	30	6
×		4
		24
+		120
		144

6.

	90	7
×		9
		63
+		810
		873

7.

	20	4
×		8
		32
+		160
		192

8.

	40	9
×		8
		72
+		320
		392

Lesson 40—Homework

PART 1

1.

	60	8
×		8
		64
+		480
		544

2.

	70	6
×		7
		42
+		490
		532

3.

	40	5
×		7
		35
+		280
		315

4.

	60	3
×		8
		24
+		480
		504

5.

	20	6
×		4
		24
+		80
		104

6.

	20	9
×		7
		63
+		140
		203

7.

	20	4
×		9
		36
+		180
		216

8.

	10	9
×		8
		72
+		80
		152

PART 2

(Anything close is acceptable.)

1. about 4 cm
2. about 8 cm
3. about 7 mm
4. about 4 mm

PART 3 (Items in bold type are given in the text.)

The Starting Number	? × 10	? × 100	? × 1,000
4,000	**400 × 10**	40 × 100	**4 × 1,000**
2,000	200 × 10	**20 × 100**	2 × 1,000
7,000	700 × 10	70 × 100	7 × 1,000
6,000	600 × 10	60 × 100	6 × 1,000
5,000	500 × 10	50 × 100	5 × 1,000
3,000	300 × 10	30 × 100	3 × 1,000
10,000	1,000 × 10	100 × 100	10 × 1,000
9,000	900 × 10	90 × 100	9 × 1,000

Lesson 41—Using Meters for Estimation

Skateboard Problem: It's about 2–3 meters from A to B (assuming the skateboard is about 1 meter long). Diver is about 2 meters tall. It takes about 2 diver lengths to get from C to D. So that's about 4 meters.

Lesson 41—Homework

PART 1

1. 48
2. 48
3. 45
4. 45
5. 480
6. 480
7. 450
8. 450
9. 4,800
10. 4,800
11. 4,500
12. 4,500

PART 2

1. 4 × 10
2. 6 × 10
3. 50 × 10
4. 40 × 10
5. 60 × 10
6. 5 × 10
7. 400 × 10
8. 600 × 10
9. 500 × 10

PART 3

(b) 700 × 10
(c) 7 × 1,000
(d) 90 × 100

PART 4

1.

30	9
×	4
	36
+	120
	156

2.

80	2
×	4
	8
+	320
	328

3.

40	7
×	8
	56
+	320
	376

4.

30	4
×	8
	32
+	240
	272

5.

50	9
×	6
	54
+	300
	354

6.

70	1
×	6
	6
+	420
	426

PART 5

(b) 800 × 10
(c) 8 × 10 × 10 × 10
(d) 20 × 10 × 10
(e) 600 × 10
(f) 6 × 10 × 10 × 10

Lesson 42—The World of Design Designing an Ad for a Web Site

See individual student work. Use the following as a guideline:

1. The stars are about $4\frac{1}{2}$ cm wide each, the title box is about $12\frac{1}{2}$ cm wide, and the description box is also about $12\frac{1}{2}$ cm wide.
2. It's about $9\frac{1}{2}$ cm between the dots in the stars.
3. The rectangle for the title of the ad is about 2 cm tall.
4. The rectangle for the description is about 5 cm tall at its tallest point.

Lesson 42—Homework

PART 1

1. 123
2. 73
3. 127

PART 2

1. 67
2. 55
3. 43

PART 3

1. 288
2. 336
3. 648

PART 4

1. Estimate: 10 × 20 = 200
   Exact Answer: 11 × 24 = 264
2. Add the figures for 1999, 2000, and 2001. The sum is $2,552,754. This is the total for the 3 years. Subtract the 1999 figure from the 2001 figure. The difference is $791,013.

Lesson 43—Making a Logo

See individual student work. This activity is designed to assess students' measuring abilities. The design needs to be an *exact* reproduction with *exact* lengths. Check to see that students have measured accurately.

Lesson 43—Homework

PART 1

1. 25
2. 72
3. 250
4. 720
5. 2,500
6. 7,200
7. 42
8. 32
9. 420
10. 320
11. 4,200
12. 3,200

PART 2 (Items in bold type are given in the text.)

The Number	? × 10	? × 100	? × 1,000
300	**30 × 10**	**3 × 100**	**nothing**
200	20 × 10	2 × 100	nothing
100	10 × 10	1 × 100	nothing
3,000	300 × 10	30 × 100	3 × 1,000
9,000	900 × 10	90 × 100	9 × 1,000

PART 3

1. Estimate: $70 \times 5 = 350$
   Exact Answer: 335
2. Estimate: $700 \times 5 = 3{,}500$
   Exact Answer: 3,335
3. Estimate: $50 \times 7 = 350$
   Exact Answer: 315
4. Estimate: $500 \times 7 = 3{,}500$
   Exact Answer: 3,465
5. Estimate: $100 \times 6 = 600$
   Exact Answer: 588
6. Estimate: $1{,}000 \times 6 = 6{,}000$
   Exact Answer: 5,928
7. Estimate: $20 \times 3 = 60$
   Exact Answer: 63
8. Estimate: $250 \times 3 = 750$
   Exact Answer: 723

Lesson 44—Homework

PART 1

1. 56
2. 54
3. 20
4. 560
5. 540
6. 200
7. 5,600
8. 5,400
9. 2,000
10. 64
11. 28
12. 200
13. 640
14. 280
15. 2,000

PART 2

1. 50 mm or 5 cm
2. 40 cm or 4 dm
3. 100 dm or 10 m
4. 60 mm or 6 cm
5. 80 cm or 8 dm

PART 3

1. The correct answer is 245. The student carried the 5 instead of the 4.
2. The correct answer is 624. The student forgot to add the carry.
3. The correct answer is 392. The student carried the 2 instead of the 4.
4. This problem does not have an error.

Lesson 45—Creating a Web Page

See individual student work. This activity is designed to assess students' measuring abilities. The design needs to be a *fairly accurate* reproduction with *fairly accurate* lengths. Be sure that the squares are 2 cm high and 2 cm wide. Also, be sure the students have followed the directions carefully. For example, the name "The Mad Hatter" should be in the top box.

Lesson 45—Homework

PART 1

1.	48	480	4,800
2.	63	630	6,300
3.	20	200	2,000
4.	27	270	2,700
5.	48	480	4,800

PART 2

1.

	60	2
×		3
		6
+		180
		186

2.

	40	1
×		5
		5
+		200
		205

3.

	40	8
×		5
		40
+		200
		240

4.

	30	2
×		7
		14
+		210
		224

5.

	20	9
×		7
		63
+		140
		203

6.

	20	5
×		4
		20
+		80
		100

7.

	90	9
×		9
		81
+		810
		891

8.

	60	2
×		4
		8
+		240
		248

Lesson 46—Problem Solving

1. 6 high and 9 wide
2. 40 million $\times$ 3 = 120 million
   80 million $\times$ 2 = 160 million

Lesson 46—Homework
PART 1

1. 54 540 5,400
2. 56 560 5,600
3. 18 180 1,800

PART 2

1.
60 | 5
× | 7
35
\+ 420
455

2.
20 | 5
× | 8
40
\+ 160
200

3.
40 | 4
× | 4
16
\+ 160
176

4.
30 | 7
× | 3
21
\+ 90
111

PART 3
Problem 3 is incorrect.
Estimate: 40 × 20 = 800
Exact Answer: 756

Lesson 47—More About Design: Creating Another Logo
See individual student work. This activity is designed to assess students' measuring abilities. The shapes need to be measured using *exact* lengths. The sides of the smallest triangle should be 2 cm, the sides of the medium triangles should be 4 cm, and the sides of the largest triangles should be 6 cm.

Lesson 47—Homework
PART 1

1. 48
2. 63
3. 480
4. 630
5. 4,800
6. 6,300
7. 48 × 100 = 4,800
8. 63 × 100 = 6,300

PART 2 (Items in bold type are given in the text.)

The Number	? × 10	? × 100	? × 1,000
5,000	**500 × 10**	**50 × 100**	**5 × 1,000**
2,000	200 × 10	20 × 100	2 × 1,000
3,000	300 × 10	30 × 100	3 × 1,000
8,000	800 × 10	80 × 100	8 × 1,000
19,000	1,900 × 10	190 × 100	19 × 1,000

PART 3

1. 4,200
2. 4,000
3. 2,400
4. 35,000
5. 3,600
6. 1,200
7. 1,600
8. 2,000

Lesson 48—Scale Drawings
See individual student answers. Use the following answers as a guideline but remember that these are approximate measures. Students can be off by about 3 mm each way and/or 1 cm each way.

Building A: about 33 mm
Building B: about 60 mm
Building C: about 52 mm
Building D: about 73 mm
Building E: about 46 mm
Building F: about 7 cm
Building G: about 4 cm
Building H: about 7 cm

Lesson 48—Homework
PART 1

1. 63
2. 18
3. 35
4. 630
5. 180
6. 350
7. 6,300
8. 1,800
9. 3,500
10. 32
11. 8
12. 320
13. 320
14. 80
15. 2,000

PART 2

1. 20 mm or 2 cm
2. 30 cm or 3 dm
3. 50 dm or 5 m
4. 100 mm or 10 cm or 1 dm
5. 100 cm or 10 dm or 1 m

PART 3

1. Estimate: 700 × 50 = 35,000
   Calculator Answer: 33,761
2. Estimate: 400 × 20 = 8,000
   Calculator Answer: 9,177
3. Estimate: 800 × 200 = 160,000
   800 × 20 = 16,000
   Calculator Answer: 15,181
4. Estimate: 200 × 500 = 100,000
   200 × 50 = 10,000
   Calculator Answer: 9,964

Lesson 49—Problem Solving

1. The total sales for Aug. through Jan. is $115,000.
2. $5,000 more sold in Nov. than in Sept.
3. 5 × ____ = 25. Answer: 5 × bigger
4. 10 × ____ = 30. Answer: 3 × bigger
5. 15 × ____ = 30. Answer: Twice as big

Lesson 49—Homework

PART 1

1. 28
2. 280
3. 54
4. 5,400
5. 63
6. 6,300
7. 63 × 100 = 6,300
8. 54 × 100 = 5,400

PART 2 (Items in bold type are given in the text.)

The Number	? × 10	? × 100	? × 1,000
4,000	**400 × 10**	**40 × 100**	**4 × 1,000**
6,000	600 × 10	60 × 100	6 × 1,000
7,000	700 × 10	70 × 100	7 × 1,000
9,000	900 × 10	90 × 100	9 × 1,000
36,000	3,600 × 10	360 × 100	36 × 1,000

PART 3

1. 2,535
2. 351,200
3. 24,000
4. 1,620
5. 5,400
6. 1,284
7. 126,720
8. 3,200

Lesson 50—Homework

PART 1

1. 42
2. 42
3. 27
4. 27
5. 420
6. 420
7. 270
8. 270
9. 4,200
10. 4,200
11. 2,700
12. 2,700

PART 2

1. 6 × 10
2. 3 × 10
3. 4 × 10
4. 8 × 10
5. 7 × 10
6. 40 × 10
7. 80 × 10
8. 70 × 10
9. 400 × 10

PART 3

1. 3 × 100
2. 9 × 1,000
3. 6 × 100
4. 7 × 1,000
5. 2 × 10
6. 2 × 10,000
7. 4 × 10
8. 5 × 100
9. 18 × 1,000

PART 4

1.

	40	2
×		4
		8
+		160
		168

2.

	60	6
×		4
		24
+		240
		264

3.

	70	1
×		8
		8
+		560
		568

4.

	50	1
×		8
		8
+		400
		408

5.

	40	9
×		6
		54
+		240
		294

6.

	90	2
×		6
		12
+		540
		552

PART 5 (Items in bold type are given in the text.)

The Number	? × 10	? × 100	? × 10 × 10
4,000	**400 × 10**	**40 × 100**	**40 × 10 × 10**
2,000	200 × 10	20 × 100	20 × 10 × 10
1,000	100 × 10	10 × 100	10 × 10 × 10
3,000	300 × 10	30 × 100	30 × 10 × 10
9,000	900 × 10	90 × 100	90 × 10 × 10

PART 6

1. 185
2. 1,029
3. 35,952

PART 7

1. The correct answer is 464. The student forgot to add the carry.
2. The correct answer is 2,747. The student forgot the place holder in the second line of traditional multiplication.
3. The correct answer is 270,599. The student forgot the place value holders in the second and third lines of the traditional multiplication.

PART 8 (Items in bold type are given in the text.)

Extended Fact	Pulling Out the 10s	Show as a Basic Fact and a Power of 10	Then Combine	And Solve
50 × 60	**5 × 10 × 6 × 10**	**5 × 6 × 10 × 10**	**30 × 100**	**3,000**
70 × 80	7 × 10 × 8 × 10	7 × 8 × 10 × 10	56 × 100	5,600
90 × 40	9 × 10 × 4 × 10	9 × 4 × 10 × 10	36 × 100	3,600
50 × 50	5 × 10 × 5 × 10	5 × 5 × 10 × 10	25 × 100	2,500
60 × 10	6 × 10 × 1 × 10	6 × 1 × 10 × 10	6 × 100	600
80 × 90	8 × 10 × 9 × 10	8 × 9 × 10 × 10	72 × 100	7,200
30 × 20	3 × 10 × 2 × 10	3 × 2 × 10 × 10	6 × 100	600

UNIT 4

STUDENT TEXTBOOK ANSWER KEY

Division

Lesson 51—Homework

PART 1

1. $9 + 5 = 14$
   $5 + 9 = 14$
   $14 - 9 = 5$
   $14 - 5 = 9$
2. $8 + 3 = 11$
   $3 + 8 = 11$
   $11 - 3 = 8$
   $11 - 8 = 3$
3. $5 + 7 = 12$
   $7 + 5 = 12$
   $12 - 7 = 5$
   $12 - 5 = 7$

PART 2

1. 7
2. 9
3. 8
4. 8
5. 9
6. 9
7. 6
8. 7
9. 9

PART 3

1. $9 \times 5 = 45$
   $5 \times 9 = 45$
   $45 \div 5 = 9$
   $45 \div 9 = 5$
2. $8 \times 3 = 24$
   $3 \times 8 = 24$
   $24 \div 8 = 3$
   $24 \div 3 = 8$
3. $6 \times 7 = 42$
   $7 \times 6 = 42$
   $42 \div 6 = 7$
   $42 \div 7 = 6$

Lesson 52—Homework

PART 1

1. 72
2. 64
3. 3
4. 30
5. 12
6. 7
7. 27
8. 7
9. 4
10. 20
11. 9
12. 8

PART 2

1.	63	630	6,300
2.	48	480	4,800
3.	8	80	800
4.	6	60	600

PART 3

1. $9 \times 5 = 45$
   $5 \times 9 = 45$
   $45 \div 9 = 5$
   $45 \div 5 = 9$
2. $8 \times 3 = 24$
   $3 \times 8 = 24$
   $24 \div 8 = 3$
   $24 \div 3 = 8$
3. $6 \times 7 = 42$
   $7 \times 6 = 42$
   $42 \div 6 = 7$
   $42 \div 7 = 6$

Lesson 53—Profits From May Concerts

1. $21,860 \div 6 =$ about \$3,645 each
2. Attendance for all 6 concerts combined was about 4,200. $4,200 \div 6$ concerts $=$ about 700 per concert average.

Lesson 53—Homework

PART 1

1. 56
2. 36
3. 12
4. 6
5. 560
6. 360
7. 120
8. 60
9. 70
10. 3,600
11. 1,200
12. 600
13. 56
14. 36
15. 12
16. 6

PART 2

See individual student work to make sure number lines are drawn correctly.

Answers are as follows:

1. 4
2. 3
3. 3
4. 4

PART 3

1. 3
2. 6
3. 8
4. 9
5. 9
6. 8

Lesson 54—Homework

PART 1

1. 24
2. 20
3. 18
4. 9
5. 240
6. 200
7. 180
8. 90
9. 30
10. 2,000
11. 1,800
12. 900
13. 3
14. 20
15. 18
16. 9

PART 2

See individual student work to make sure number lines are drawn correctly.

Answers are as follows:

1. 3
2. 9
3. 3
4. 5

PART 3

1. 20
2. 3
3. 40
4. 9
5. 40
6. 7

Lesson 55—Homework

PART 1

See individual student work to make sure the number lines are drawn correctly.

Answers are as follows:

1. 30
2. 50
3. 40
4. 60

PART 2

1. Box the 7 and circle the .75.
2. Box the 3 and circle the .5.
3. Box the 8 and circle the .5.
4. Box the 4 and circle the .5.

PART 3

1. 7 R4
2. 3 R3
3. 8 R1
4. 4 R4

Lesson 56—Problem Solving

1. About 19 boxes each
2. Most: Thursday
   Least: Friday
   Total: 24 hours
3. Shows on Tuesday night: 140
   Difference sports vs. news: 13
   Movies, reruns, and dramas: 84
   Sports and talk shows: 37
   Difference between Rene's shows and Carina's shows: 47
   Game shows on Saturday: 17
   Articles on the Internet each day:
   $8 \times 4 = 32$ million
4. If there are the same number each day and she watches for 7 days (Thursday – Wednesday), then there are $119 \div 7$ or 17 each day. Therefore, there are 17 on Saturday.
5. If there are 8 million people who read 4 articles per day, then there are 8×4 or 32 million articles read on the Internet every day.

Lesson 56—Homework

PART 1

1. 8
2. 70
3. 80
4. 80

PART 2

1. 9
2. 6
3. 10
4. 6
5. 7
6. 3

Lesson 57—Homework

PART 1

1. 21
2. 36
3. 48
4. 7
5. 210
6. 360
7. 480
8. 70
9. 30
10. 3,600
11. 4,800
12. 700
13. 21
14. 36
15. 48
16. 7

PART 2

1. $72 \div 9 = 8$
   Calculator Answer: 8.33
2. $49 \div 7 = 7$
   Calculator Answer: 7.29
3. $24 \div 3 = 8$
   Calculator Answer: 7.67
4. $36 \div 4 = 9$
   Calculator Answer: 9.5
5. $64 \div 8 = 8$
   Calculator Answer: 8.25
6. $36 \div 6 = 6$
   Calculator Answer: 5.67
7. $16 \div 4 = 4$
   Calculator Answer: 4
8. $45 \div 5 = 9$
   Calculator Answer: 9.4

Lesson 58—Homework

PART 1

1. 7
2. 9
3. 7
4. 3
5. 8
6. 7
7. 7
8. 9

PART 2

1. 70
2. 900
3. 70
4. 300
5. 80
6. 700
7. 70
8. 900

PART 3

1. $7 \times 8 = 56$
   $8 \times 7 = 56$
   $56 \div 8 = 7$
   $56 \div 7 = 8$
2. $9 \times 2 = 18$
   $2 \times 9 = 18$
   $18 \div 2 = 9$
   $18 \div 9 = 2$
3. $4 \times 6 = 24$
   $6 \times 4 = 24$
   $24 \div 4 = 6$
   $24 \div 6 = 4$
4. $4 \times 3 = 12$
   $3 \times 4 = 12$
   $12 \div 3 = 4$
   $12 \div 4 = 3$
5. $5 \times 6 = 30$
   $6 \times 5 = 30$
   $30 \div 6 = 5$
   $30 \div 5 = 6$
6. $4 \times 9 = 36$
   $9 \times 4 = 36$
   $36 \div 4 = 9$
   $36 \div 9 = 4$

PART 4

See individual student work to make sure number lines are drawn correctly.

Answers are as follows:

1. 3
2. 4
3. 20
4. 5
5. 30
6. 70

PART 5

1. 7
   Calculator Answer: 7.25
2. 5
   Calculator Answer: 4.5
3. 9
   Calculator Answer: 8.5
4. 3
   Calculator Answer: 2.5

PART 6

1. $63 \div 9 = 7$
2. $42 \div 7 = 6$
3. $48 \div 8 = 6$
4. $18 \div 6 = 3$

Lesson 59—Homework

PART 1

1. 5
2. 3
3. 30
4. 6
5. 40
6. 50

PART 2

1. 43
2. 67
3. 79
4. 59

PART 3

$1,472 \div 4 = \$368$

Lesson 60—Homework

PART 1

1. 5
2. 40
3. 60
4. 90

PART 2

1. $27 \div 9 = 3$
2. $21 \div 7 = 3$
3. $16 \div 8 = 2$
4. $30 \div 6 = 5$

PART 3

1. $270 \div 9 = 30$
2. $210 \div 7 = 30$
3. $160 \div 8 = 20$
4. $300 \div 6 = 50$

Lesson 61—Homework

PART 1

1. $54 \div 9 = 6$
2. $49 \div 7 = 7$
3. $64 \div 8 = 8$
4. $35 \div 5 = 7$

PART 2

1. $560 \div 80 = 7$
2. $490 \div 70 = 7$
3. $320 \div 80 = 4$
4. $280 \div 40 = 7$

PART 3

1. $9 \times 10\overline{)36 \times 10}$
2. $7 \times 10\overline{)49 \times 10}$
3. $8 \times 10\overline{)32 \times 10}$
4. $4 \times 10\overline{)28 \times 10}$

Lesson 62—The Arena Building

1. 89 is about 90. $90 \div 10$ is about 9 floors.
2. $90 \times 2 = 180$. There are about 180 windows on one side of the building.
3. 20 tables × 5 rooms = 100 tables
4. 238 companies ÷ 80 floors ⟶ $240 \div 80 =$ about 3 companies per floor
5. There are 10 hours between 8 A.M. and 6 P.M. The total people would be 2,019 ÷ 10 hours $= 2{,}000 \div 10 =$ about 200 people per hour.

Lesson 62—Homework

PART 1

1. 6	60	600
2. 4	40	400
3. 9	90	900
4. 2	20	200
5. 6	60	600

PART 2

1. $160 \div 20 = 8$
2. $360 \div 40 = 9$
3. $270 \div 30 = 9$
4. $240 \div 30 = 8$
5. $250 \div 50 = 5$
6. $720 \div 90 = 8$

PART 3

1. $9 \times 10\overline{)72 \times 10}$

$9\overline{)72}$ = 8

Answer: 8

2. $7 \times 10\overline{)42 \times 10}$

$7\overline{)42}$ = 6

Answer: 6

3. $8 \times 10\overline{)40 \times 10}$

$8\overline{)40}$ = 5

Answer: 5

4. $4 \times 10\overline{)12 \times 10}$

$4\overline{)12}$ = 3

Answer: 3

Lesson 63—Making Division Word Problems

See individual student work. Be sure that the students' problems involve division. Look for at least four sentences for each problem. Be sure that the students are able to explain why the problems are division problems. Look for discussions about "dividing something into equal parts." Then, be sure the students have answered the problems correctly.

Lesson 63—Homework

PART 1

1. 8
2. 3
3. 8
4. 7
5. 6
6. 3
7. 9
8. 5
9. 2
10. 6
11. 6
12. 4

PART 2

1. $240 \div 30 = 8$
2. $250 \div 50 = 5$
3. $360 \div 60 = 6$
4. $640 \div 80 = 8$

PART 3

1. $3 \times 10\overline{)18 \times 10}$

$3\overline{)18}$ = 6

Answer: 6

2. $9 \times 10\overline{)36 \times 10}$

$9\overline{)36}$ = 4

Answer: 4

3. $5 \times 10\overline{)25 \times 10}$

$5\overline{)25}$ = 5

Answer: 5

4. $7 \times 10\overline{)49 \times 10}$

$7\overline{)49}$ = 7

Answer: 7

PART 4

1. Estimate: $420 \div 7 = 60$
2. Calculator Answer: $429 \div 7 = 61.29$
3. It looks like the student entered the problem in backward, dividing the smaller number by the bigger number.

Lesson 64—Homework

PART 1

1. 7
2. 3
3. 7
4. 7
5. 60
6. 5
7. 9
8. 6
9. 20
10. 60
11. 90
12. 2

PART 2

1. $630 \div 90 = 7$
2. $560 \div 70 = 8$
3. $400 \div 50 = 8$
4. $480 \div 80 = 6$

PART 3

1. Estimate: $180 \div 20 = 9$
2. Calculator Answer: 8.43
3. It looks like the student entered the digits wrong, maybe entering $149 \div 23$. That's about $140 \div 20$, and that's close to 7. When I try that on my calculator, I get 6.478. That's it!

Lesson 65—Homework

PART 1

1. 7
2. 90
3. 7
4. 30
5. 8
6. 70
7. 7
8. 90

PART 2

1. $7 \times 8 = 56$
   $8 \times 7 = 56$
   $56 \div 7 = 8$
   $56 \div 8 = 7$
2. $9 \times 8 = 72$
   $8 \times 9 = 72$
   $72 \div 9 = 8$
   $72 \div 8 = 9$
3. $4 \times 6 = 24$
   $6 \times 4 = 24$
   $24 \div 4 = 6$
   $24 \div 6 = 4$
4. $5 \times 6 = 30$
   $6 \times 5 = 30$
   $30 \div 6 = 5$
   $30 \div 5 = 6$

PART 3

See individual student work to be sure number lines are drawn correctly.

Answers are as follows:

1. 3
2. 4

PART 4

1. $81 \div 9 = 9$
2. $64 \div 8 = 8$

PART 5

1. 52
2. 43

PART 6

1. $270 \div 9 = 30$
2. $160 \div 8 = 20$

PART 7

1. $240 \div 80 = 3$
2. $150 \div 30 = 5$

PART 8

1. $9 \times 10 \overline{)45 \times 10}$
2. $8 \times 10 \overline{)72 \times 10}$

PART 9

1. Estimate: $560 \div 70 = 8$
2. Calculator Answer: 7.875
3. It looks like the student entered the numbers in backward. That is, the student divided the small number by the bigger number.

UNIT 5

STUDENT TEXTBOOK ANSWER KEY

Factors, Patterns, and Multiples

Lesson 66—Car Sales on the Internet

1. Car Trader sold about 15,000. Auto sold about 12,200. The difference is about 2,800.
2. All four companies sold about 39,200 cars in May. (12,200 + 9,000 + 3,000 + 15,000)
3. The sales were 14,500. There were 5 people. 15,000 ÷ 5 = 3,000. Tanya sold about 2,900 cars.
4. The car company that sold the most cars during May is Car Trader with 15,000. The car company that sold the least is More Cars with 3,000. The difference is about 12,000.

Lesson 66—Homework

PART 1

1. The estimate is 400 + 450 = 850. The exact answer is 845.
2. The estimate is 12,000 − 9,000 = 3,000. The exact answer is 2,915.
3. The estimate is 2,500 ÷ 50 = 50. The exact answer is 48.
4. The estimate is 200 × 7 = 1,400. The exact answer is 1,484.
5. The estimate is 90 × 6 = 540. The exact answer is 534.
6. The estimate is 2,800 ÷ 28 = 100. The exact answer is 103.77.

PART 2

1. There are 8 units. Another array could look like this:
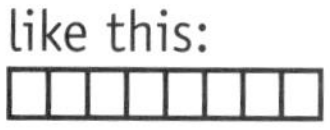
2. There are 4 units. Another array could look like this:
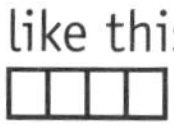
3. There are 6 units. Another array could look like this:
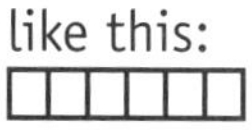
4. There are 10 units. Another array could look like this:
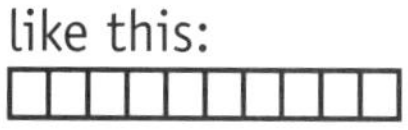
5. There are 10 units. Another array could look like this:
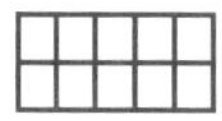

Lesson 67—Homework

PART 1

1. The estimate is 33,000 − 30,000 = 3,000. The exact answer is 3,225.
2. The estimate is 475 + 325 = 800. The exact answer is 805.
3. The estimate is 600 ÷ 30 = 20. The exact answer is 21.
4. The estimate is 1,000 × 9 = 9,000. The exact answer is 9,180.
5. The estimate is 99 ÷ 9 = 11. The exact answer is 11.4.
6. The estimate is 1,000 × 6 = 6,000. The exact answer is 5,934.

PART 2

See individual student responses. There is more than one correct answer. Just be sure the arrays represent the numbers given.

1. There are 16 units. Another array could look like this:
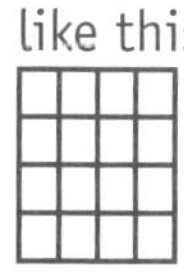
2. There are 14 units. Another array could look like this:

3. There are 12 units. Another array could look like this:

4. There are 15 units. Another array could look like this:
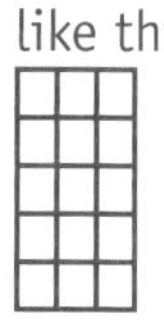

5. There are 20 units. Another array could look like this:

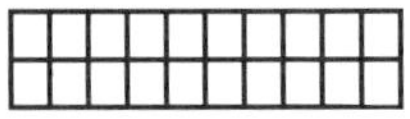

Lesson 68—Homework

PART 1

1. $2 \times 4 = 8$. 2 and 4 are the factors.
2. $1 \times 3 = 3$. 1 and 3 are the factors.
3. $2 \times 3 = 6$. 2 and 3 are the factors.
4. $1 \times 6 = 6$. 1 and 6 are the factors.

PART 2

Number	Multiplication Problems	Factors
8	$1 \times 8 = 8$ $2 \times 4 = 8$	1, 2, 4, 8
15	$1 \times 15 = 15$ $3 \times 5 = 15$	1, 3, 5, 15
7	$1 \times 7 = 7$	1, 7
25	$1 \times 25 = 25$ $5 \times 5 = 25$	1, 5, 25

Lesson 69—Homework

PART 1

1. X = 18
2. X = 15
3. Z = 11
4. Y = 32
5. Z = 2
6. X = 42
7. Z = 4
8. Y = 5
9. Y = 2

PART 2

1. The factors of 3 are 1 and 3.

2. The factors of 6 are 1, 2, 3, and 6.

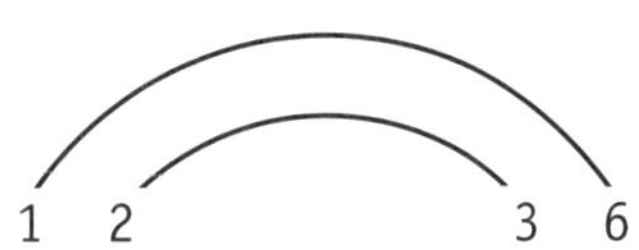

3. The factors of 9 are 1, 3, and 9.

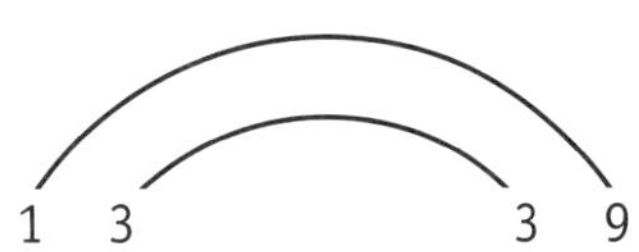

4. The factors of 24 are 1, 2, 3, 4, 6, 8, 12, and 24.

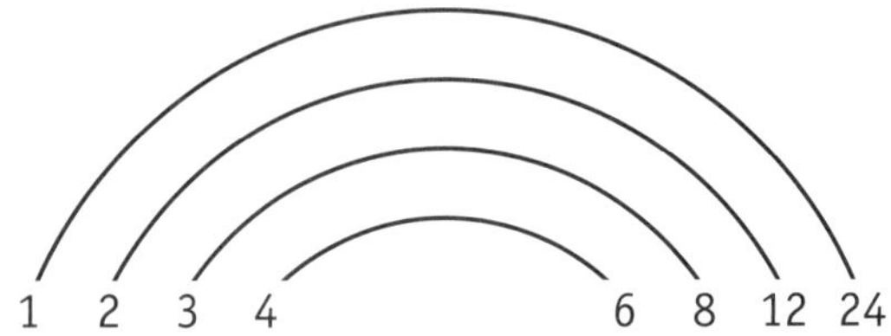

PART 3

1. $14 \times 3 = 42$ sq. units.
2. $7 \times 6 = 42$ sq. units.
3. $17 \times 1 = 17$ sq. units.
4. $11 \times 7 = 77$ sq. units.

Lesson 70—Reviewing Whole Number Operations

1. 4 walls $\times$ 12 studs = 48 studs.
2. 12 studs $\times$ 6 nails = 72 nails for one wall. $4 \times 72 = 288$ nails for 4 walls.
3. $985 - 720 = 265$. The bottom floor is 265 square units bigger than the top floor.

Lesson 70—Homework

PART 1

1. X = 18
2. X = 180
3. X = 11
4. X = 110
5. X = 2
6. X = 2
7. X = 4
8. X = 4
9. X = 40

PART 2

1. 1, 7
2. 1, 2, 3, 4, 6, and 12
3. 1, 2, 7, and 14
4. 1, 2, 3, 4, 6, 8, 12, and 24

PART 3

1. 1, 3, 7, and 21
2. 1, 3, 9, and 27
3. 1, 2, 3, 4, 6, 9, 12, 18, and 36

Lesson 71—Homework

PART 1

1. 1 and 5
2. 1, 2, 7, and 14
3. 1 and 17
4. 1, 5, and 25

PART 2

1. 2
2. 3
3. 31
4. 11
5. 23
6. 37
7. 47

PART 3

1. 4
2. 20
3. 33
4. 27
5. 49
6. 87
7. 39

Lesson 72—Where Are the Primes?

1. See individual student work. The graph should look something like this:

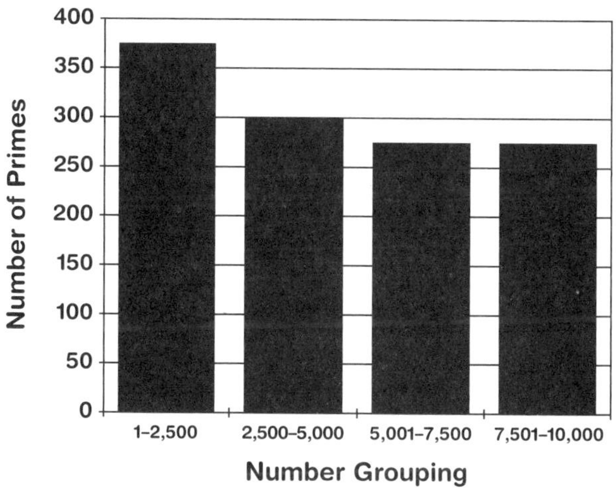

2. See individual student work. The students should include some discussion about the number of primes decreasing as the numbers get larger. The higher the numbers, the fewer primes there are.

Lesson 72—Homework

PART 1

1. 1 and 7
2. 1, 2, 4, 8, and 16
3. 1, 3, 7, and 21
4. 1, 2, 4, 8, 16, and 32

PART 2

1. 4
2. 9
3. 15
4. 25

PART 3

1. A has the biggest area (A = 99, B = 90, and C = 91).
2. A is 9 squares bigger than B.

Lesson 73—Homework

PART 1

1. Estimate is 12,400 + 1,400 = 13,800.
   Exact answer is 13,845.
2. Estimate is 700 × 3 = 2,100.
   Exact answer is 2,103.
3. Estimate is 300 ÷ 6 = 50.
   Exact answer is 50 R2.
4. Estimate is 17,000 − 10,000 = 7,000.
   Exact answer is 6,991.
5. Estimate is 100 × 8 = 800.
   Exact answer is 792.
6. Estimate is 500 ÷ 5 = 100.
   Exact answer is 99 R4.

PART 2

Factors of 22 are 1, 2, 11, and 22.
Factors of 6 are 1, 2, 3, and 6.
Factors of 30 are 1, 2, 3, 5, 6, 10, 15, and 30.
Factors of 13 are 1 and 13.
Factors of 4 are 1, 2, and 4.

PART 3

1. B has the biggest perimeter.
2. The perimeter of B is 4 units bigger than the perimeter of C.

Lesson 74—Homework

PART 1

The numbers that are not prime are 1, 4, 30, 100, and 1,000.

PART 2

The factors of 14 are 1, 2, 7, and 14.
The factors of 16 are 1, 2, 4, 8, and 16.
The factors of 17 are 1 and 17.
The factors of 24 are 1, 2, 3, 4, 6, 8, 12, and 24.
The factors of 28 are 1, 2, 4, 7, 14, and 28.

PART 3

1. a = 2, b = 2, c = 2
2. a = 3, b = 3
3. a = 2, b = 2, c = 3, d = 3
4. a = 2, b = 3 (or the reverse), and c = 7
5. a = 7, b = 7

Lesson 75—Homework

PART 1

The factors of 10 are 1, 2, 5, and 10.
The factors of 20 are 1, 2, 4, 5, 10, and 20.
The factors of 30 are 1, 2, 3, 5, 6, 10, 15, and 30.
The factors of 40 are 1, 2, 4, 5, 8, 10, 20, and 40.
The factors of 50 are 1, 2, 5, 10, 25, and 50.

PART 2

1. a = 2, b = 2, c = 3, d = 3
2. a = 2 and b = 5 (or the reverse)
3. a = 2 and b = 5 (or the reverse)
4. a = 2 and b = 5 (or the reverse) and c = 2 and d = 3 (or the reverse)
5. a = 3, b = 3, c = 5
6. a = 5, b = 5

PART 3

The perimeter is 3 + 6 + 6 + 9 + 9 + 15 = 48 cm.

Lesson 76—Homework

PART 1 (Items in bold type are given in the text.)

Number	10s	100s	1,000s
7,000	**700 × 10**	**70 × 100**	**7 × 1,000**
80	8 × 10		
90	9 × 10		
700	70 × 10	7 × 100	
500	50 × 10	5 × 100	
6,000	600 × 10	60 × 100	6 × 1,000
9,000	900 × 10	90 × 100	9 × 1,000
5,000	500 × 10	50 × 100	5 × 1,000
8,000	800 × 10	80 × 100	8 × 1,000

PART 2

Students' trees may differ depending on the facts they used, but the prime factors will be the same.

1.

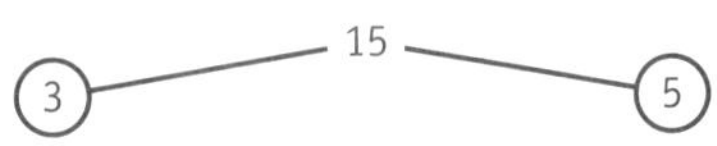

2.

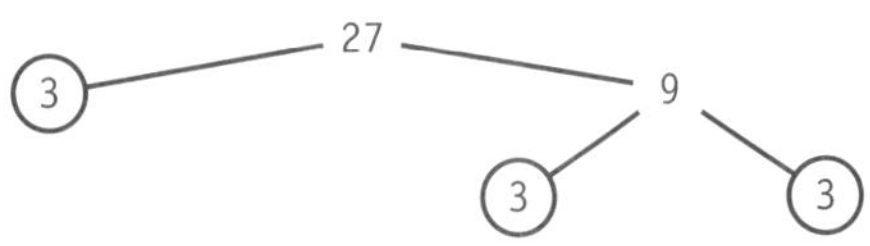

3.

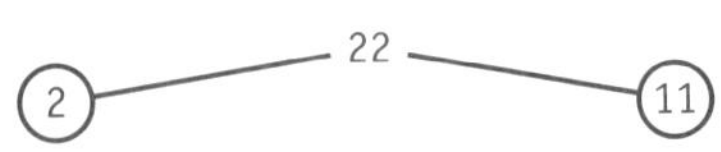

4.

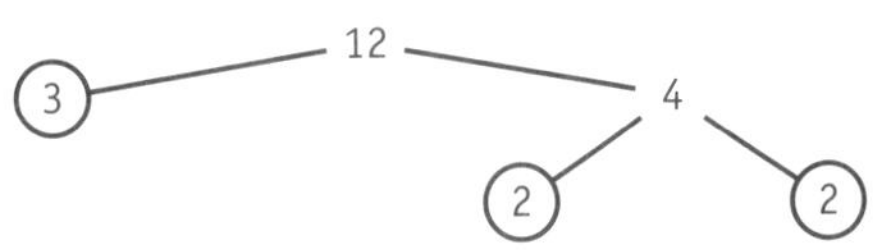

PART 3

1. Highest game was about 51 points, and the lowest game was about 28 points. The difference is 23 points.
2. In her first three games, Sarah had about 105 points. In her last three games, Sarah had about 122 points. She had more points in her last three games.
3. Sarah had about 32 points in Game 5. Her teammates had about 96 points (32 × 3). That is a total of 128 points (96 + 32 = 128).

Lesson 77—Homework

PART 1

1. The estimate is 1,700 + 2,400 = 4,100. The exact answer is 4,140.
2. The estimate is 500 + 400 + 1,000 = 1,900. The exact answer is 1,897.
3. The estimate is 400 ÷ 4 = 100. The exact answer is 102 R2.
4. The estimate is 1,000 × 5 = 5,000. The exact answer is 4,995.
5. The estimate is 600 × 6 = 3,600. The exact answer is 3,612.
6. The estimate is 500 ÷ 5 = 100. The exact answer is 99 R2.

PART 2

The factors of 10 are 1, 2, 5, and 10.
The factors of 12 are 1, 2, 3, 4, 6, and 12.
The factors of 14 are 1, 2, 7, and 14.
The factors of 16 are 1, 2, 4, 8, and 16.
The factors of 18 are 1, 2, 3, 6, 9, and 18.

PART 3

Students' trees may differ depending on the facts they used, but the prime factors will be the same.

1.

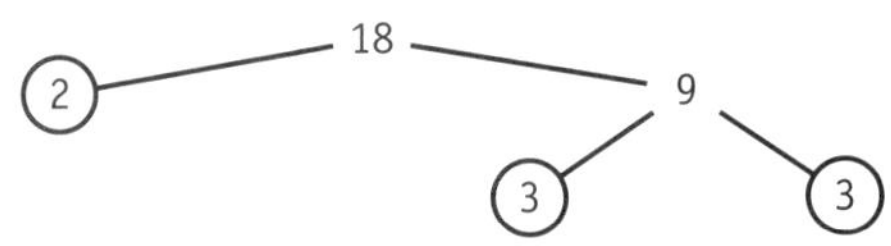

2.

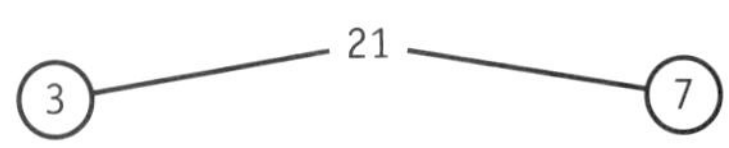

3.

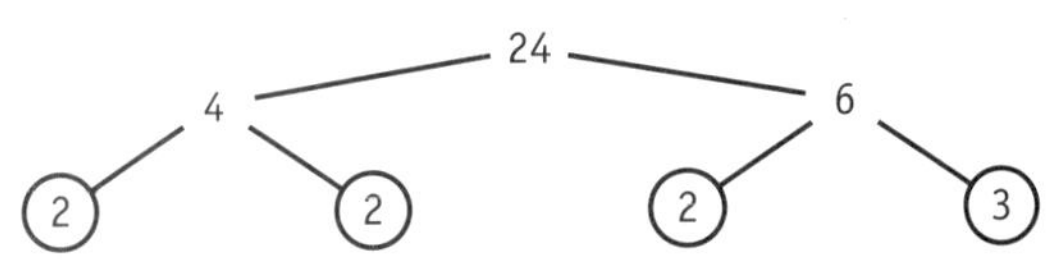

Lesson 78—Homework

PART 1

1. 12
   1 × 12

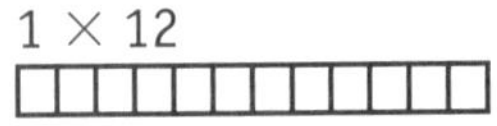

2×6

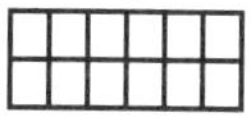

3×4

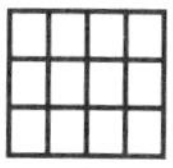

2. 7

1×7

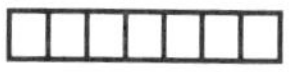

3. 16

1×16

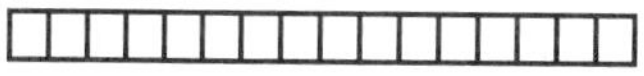

2×8

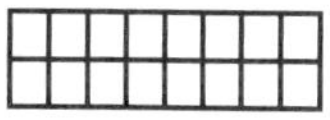

4×4

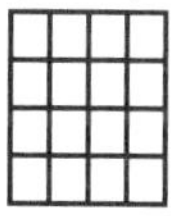

4. 21

1×21

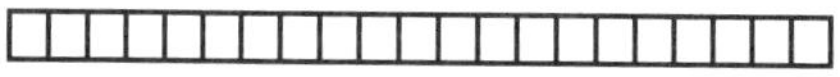

3×7

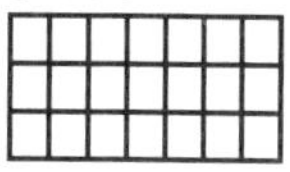

PART 2

Factors of 12 are 1, 2, 3, 4, 6, and 12.
Factors of 7 are 1 and 7.
Factors of 16 are 1, 2, 4, 8, and 16.
Factors of 21 are 1, 3, 7, and 21.

PART 3

1.

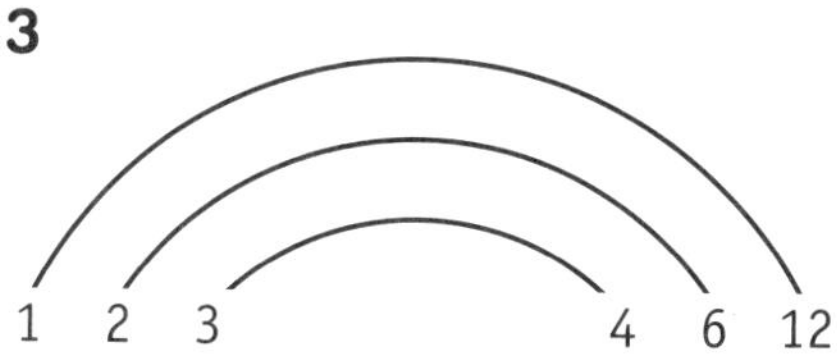

The factors of 12 are 1, 2, 3, 4, 6, and 12.

2.

The factors of 7 are 1 and 7.

3.

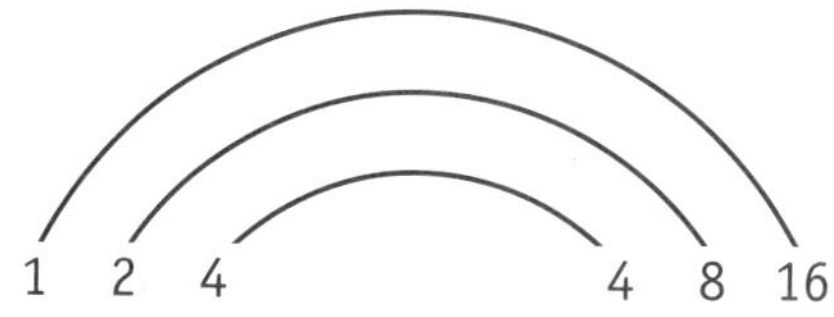

The factors of 16 are 1, 2, 4, 8, and 16.

4.

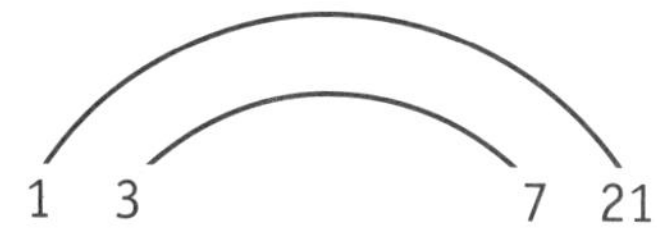

The factors of 21 are 1, 3, 7, and 21.

PART 4

1. b
2. a
3. b
4. a

PART 5

Students' trees may differ depending on the facts they used, but the prime factors will be the same.

1.

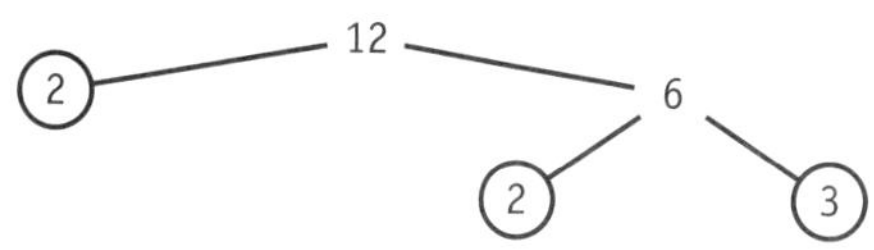

2.

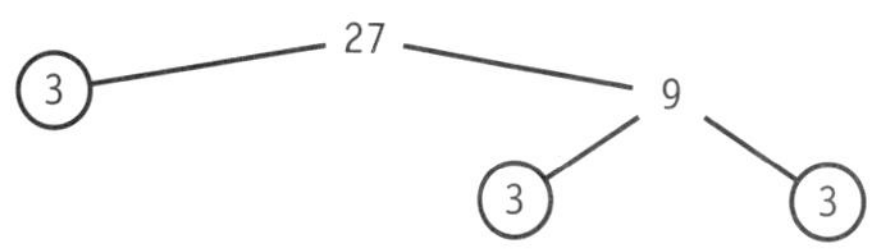

3.

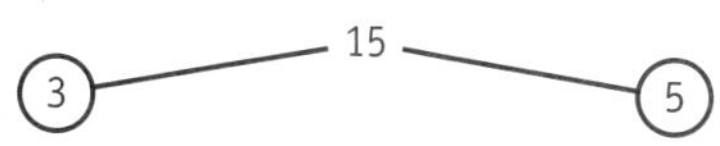

4.

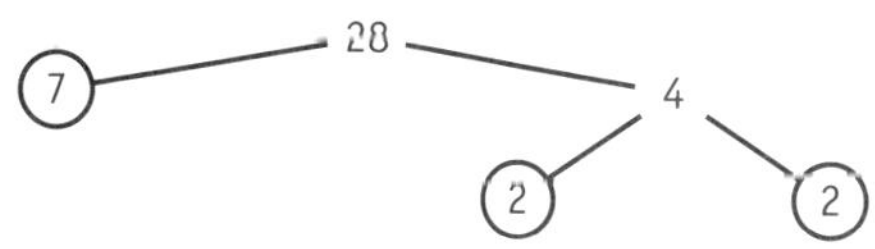

PART 6

1. Area is $2 \times 8 = 16$ sq. cm.
Perimeter is $2 + 2 + 8 + 8 = 20$ cm.
2. Area is $4 \times 4 = 16$ sq. cm.
Perimeter is $4 + 4 + 4 + 4 = 16$ cm.
3. Area is $6 \times 3 = 18$ sq. cm.
4. Area is $\frac{1}{2}(10 \times 2) = 10$ sq. cm.

Lesson 79—Homework

PART 1

1. 25
2. 56
3. 32
4. 27
5. 54
6. 63
7. 30
8. 24

PART 2

1. the 2 rule
2. the 3 rule
3. the 3 and 5 rules
4. the 2, 3, 5, 6, and 10 rules
5. the 2, 3, and 6 rules
6. the 2, 5, and 10 rules

PART 3

See individual student answers. Some sample answers are:

1. 4 (an even number)
2. 156 (a number whose digits add up to a multiple of 3)
3. 27,895 (anything ending in a 5)
4. 4,284 (an even number whose digits add up to a multiple of 3)
5. 49,570 (anything ending in a 0)

Lesson 80—Homework

PART 1

1. The factors of 36 are 1, 2, 3, 4, 6, 9, 12, 18, and 36.
2. The factors of 27 are 1, 3, 9, and 27.
3. The factors of 20 are 1, 2, 4, 5, 10, and 20.
4. The factors of 45 are 1, 3, 5, 9, 15, and 45.

PART 2

1. the 2 rule
2. the 2, 3, and 6 rules
3. the 2, 3, and 6 rules
4. the 2, 3, 5, 6, and 10 rules
5. the 3 and 5 rules
6. the 2, 3, 5, 6, and 10 rules

PART 3

1. a = 2, b = 2, c = 2, and d = 2
2. a = 2 and b = 3 (or the reverse) and c = 2 and d = 3 (or the reverse)
3. a = 7 and b = 7
4. a = 9 and b = 3
5. a = 3 and b = 5 (or the reverse)

PART 4

1. Both shapes have curved lines.
2. Both shapes have straight lines, or both are enclosed spaces.
3. Both shapes have straight lines, or both have a right angle.

Lesson 81—Homework

PART 1

1. The factors of 22 are 1, 2, 11, and 22.
2. The factors of 13 are 1 and 13.
3. The factors of 28 are 1, 2, 4, 7, 14, and 28.
4. The factors of 33 are 1, 3, 11, and 33.

PART 2

1. the 3 rule
2. the 2 rule
3. the 2, 3, and 6 rules
4. the 2, 5, and 10 rules
5. the 5 rule
6. the 2, 3, 5, 6, and 10 rules

PART 3

Students' trees may differ depending on the facts they used, but the prime factors will be the same.

1.

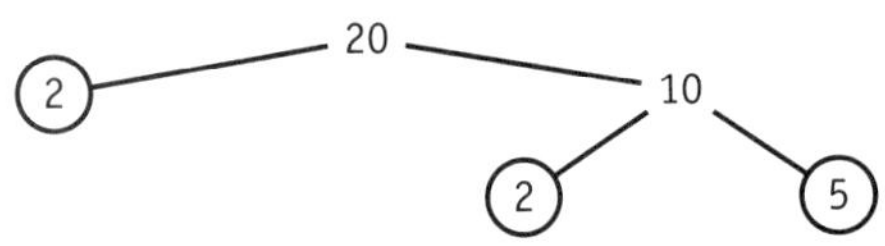

2.

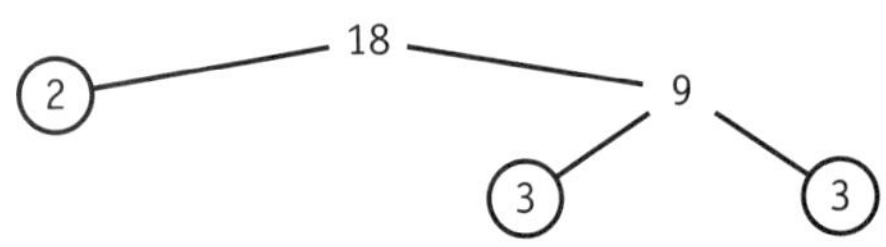

3.

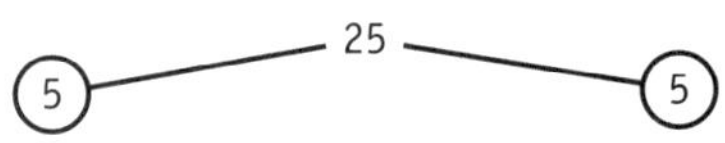

4.

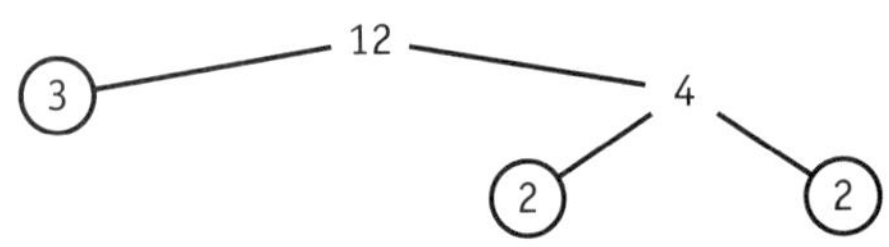

5.

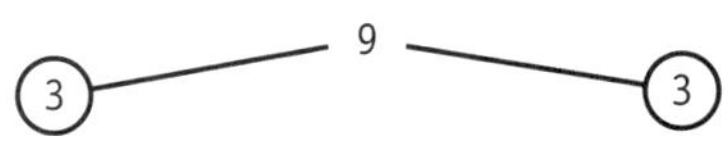

PART 4

Sample answer:

Shapes With Straight Sides	Shapes With Curved Sides	Shapes With at Least Four Corners	Shapes That Are Closed
A, B, D, and E	C	B and E	A, B, C, and E

Lesson 82—Homework

PART 1

1. The factors of 45 are 1, 3, 5, 9, 15, and 45.
2. The factors of 27 are 1, 3, 9, and 27.
3. The factors of 40 are 1, 2, 4, 5, 8, 10, 20, and 40.
4. The factors of 60 are 1, 2, 3, 4, 5, 6, 10, 12, 15, 20, 30, and 60.

PART 2

1. the 2, 3, 5, 6, and 10 rules
2. the 2, 3, and 6 rules
3. the 5 rule
4. the 2 rule
5. the 2, 3, 5, 6, and 10 rules
6. the 3 rule

PART 3

Students' trees may differ depending on the facts they used, but the prime factors will be the same.

1.
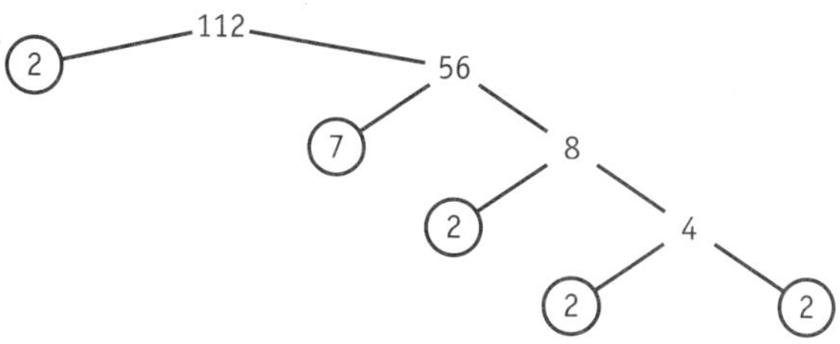

2.
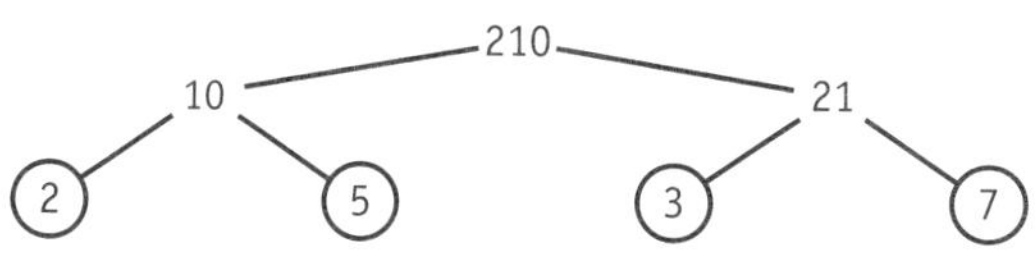

3.
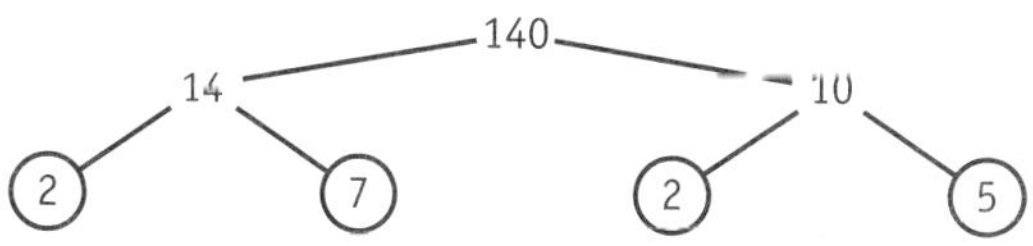

Lesson 83—Homework

PART 1

1. Common factors are 1 and 2.
2. Common factors are 1 and 2.
3. Common factors are 1 and 3.

PART 2

1. the 2 rule
2. the 3 rule
3. the 3 rule
4. the 2, 3, 5, 6, and 10 rules
5. the 2, 3, and 6 rules
6. the 5 rule

PART 3

Students' trees may differ depending on the facts they used, but the prime factors will be the same.

1.
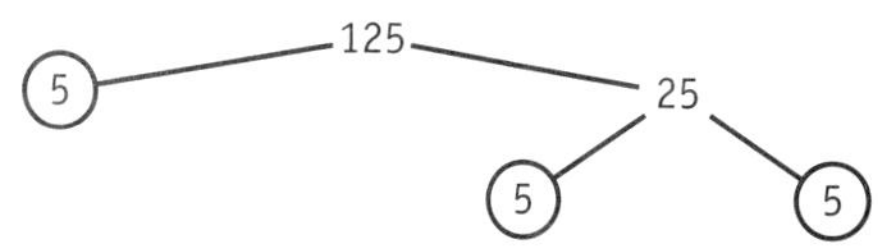

2.
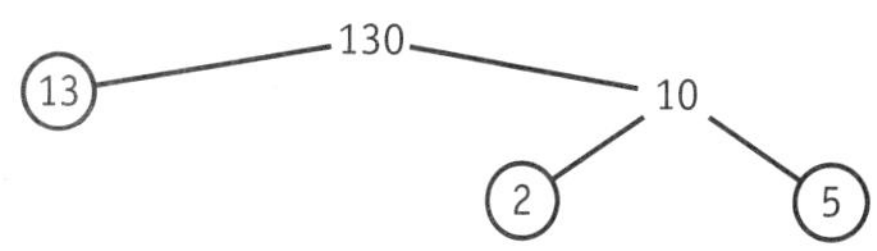

3.
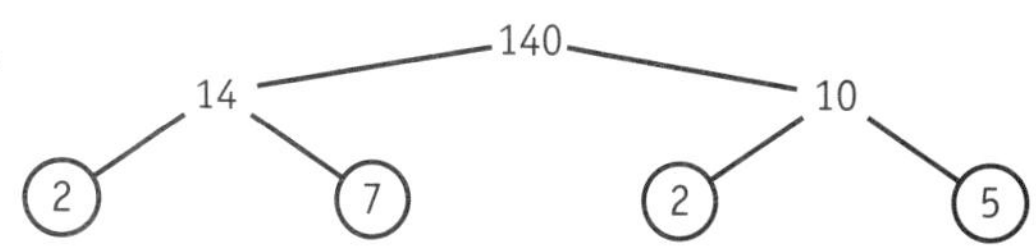

4.
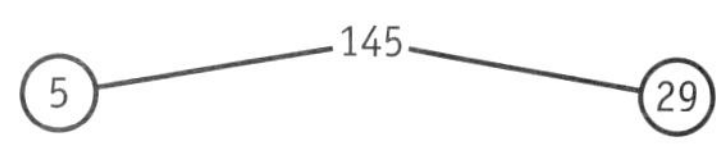

5.
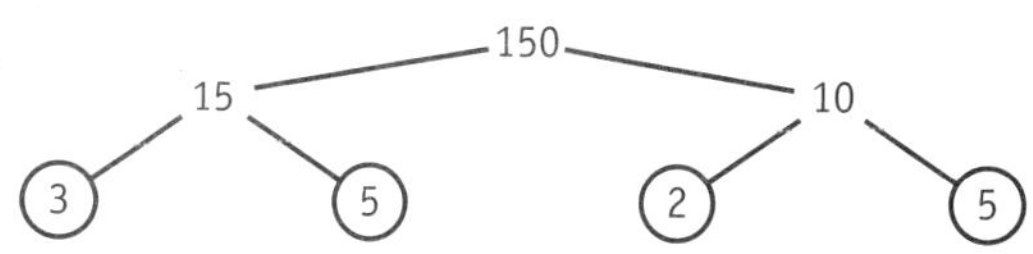

Lesson 83—Patterns in Art

See individual student responses. Answers should include some of the following discussions:

1. The designs all use a shape.
   The designs all use a lot of straight lines.
2. The first design uses a lot of squares.
   The second design uses a lot of triangles.
   The third design uses a lot of rectangles.
3. They all use geometry in their designs because they all use shapes.

Lesson 84—Patterns in Stained Glass

See individual student answers. Be sure the discussions include some of the following ideas:

1. Some of the different shapes used are trapezoids, rectangles, and triangles.
2. When all the shapes are put together, they form circular shapes with points that resemble a star or even the sun.

Lesson 84—Homework

PART 1

1. The common factor of 7 and 9 is 1.
2. The common factors of 18 and 21 are 1 and 3.
3. The common factors of 25 and 30 are 1 and 5.
4. The common factor of 32 and 37 is 1.

PART 2

1. the 2 rule
2. the 5 rule
3. the 2, 5, and 10 rules
4. the 3 and 5 rules
5. the 2, 3, and 6 rules
6. the 2, 3, and 6 rules

PART 3

1. The GCF of 18 and 21 is 3.
2. The GCF of 25 and 35 is 5.
3. The GCF of 24 and 36 is 12.
4. The GCF of 19 and 23 is 1.

Lesson 85—Homework

PART 1

1. The common factors of 20 and 25 are 1 and 5.
2. The common factor of 9 and 13 is 1.
3. The common factors of 40 and 50 are 1, 2, 5, and 10.
4. The common factor of 21 and 22 is 1.

PART 2

1. The GCF is 9.
2. The GCF is 8.
3. The GCF is 10.
4. The GCF is 7.
5. The GCF is 10.

PART 3

Students' trees may differ depending on the facts they used, but the prime factors will be the same.

1.

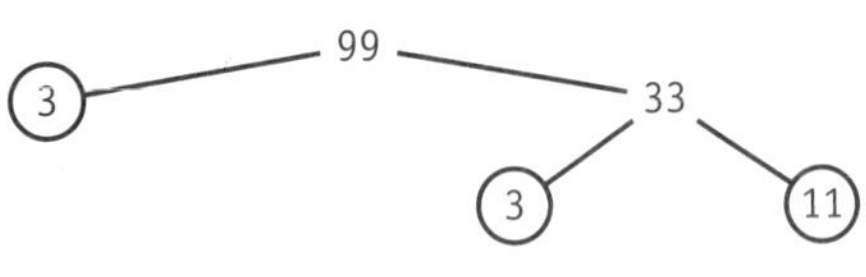

2.

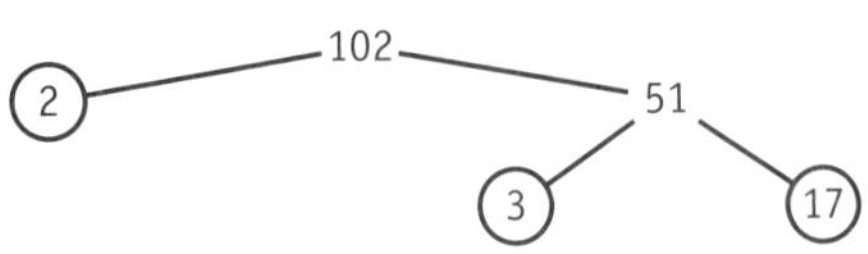

3.

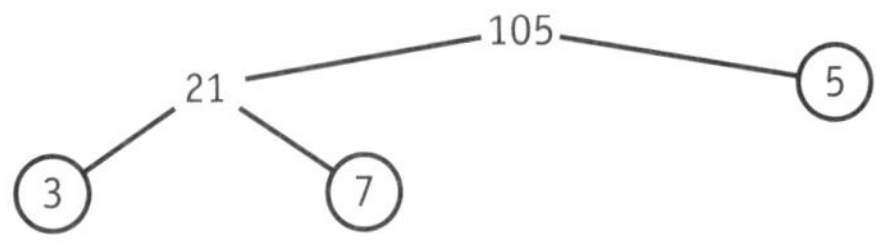

4.

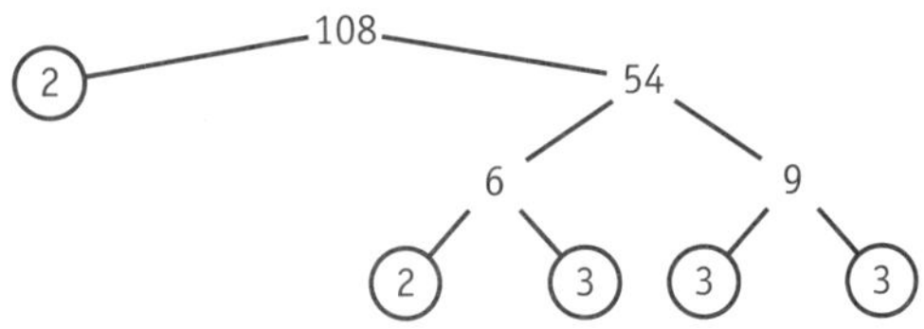

5.

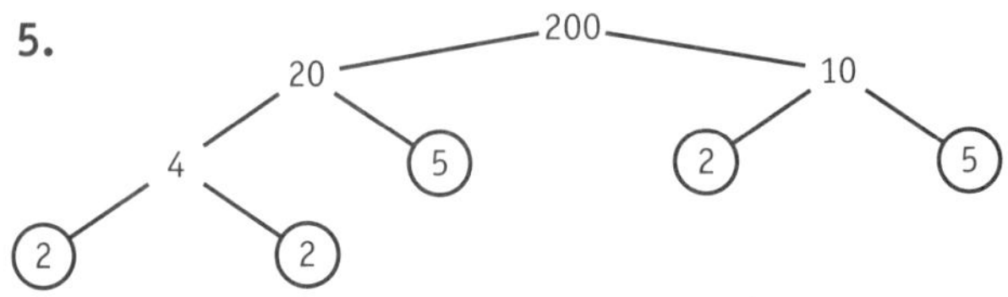

Lesson 86—Homework

PART 1

Students' trees may differ depending on the facts they used, but the prime factors will be the same.

1.

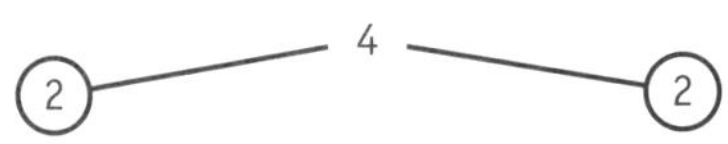

2.

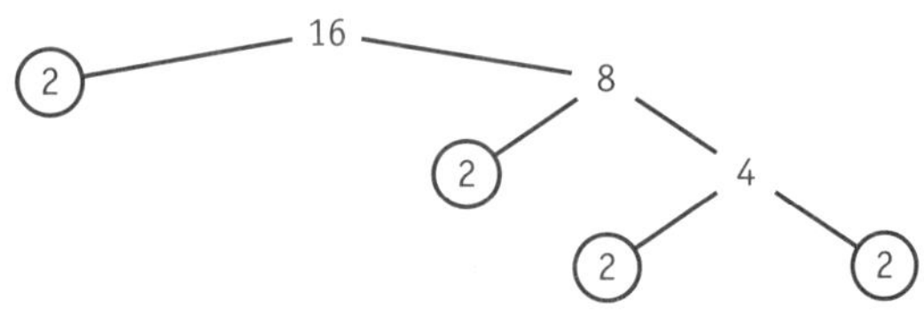

3.

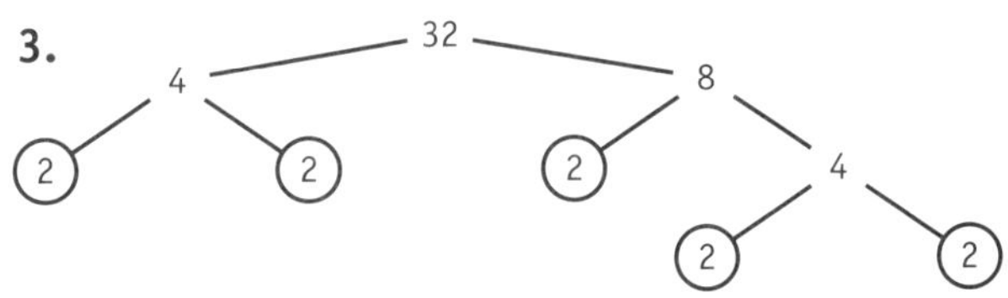

4.

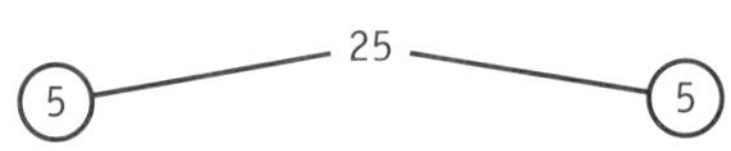

5.

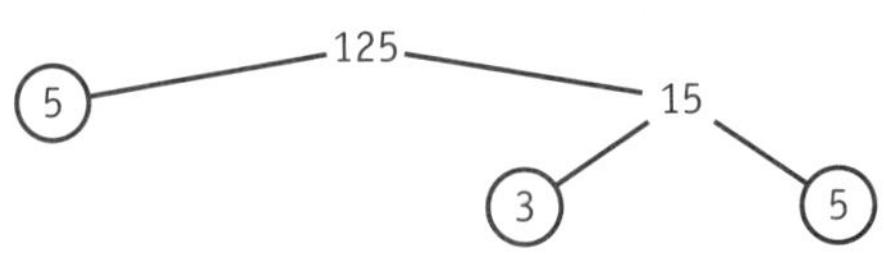

Lesson 87—Homework

PART 1

1. The common factors of 18 and 20 are 1 and 2.
2. The common factors of 20 and 22 are 1 and 2.
3. The common factors of 20 and 25 are 1 and 5.
4. The common factors of 25 and 30 are 1 and 5.

PART 2

Students' trees may differ depending on the facts they used, but the prime factors will be the same.

1.

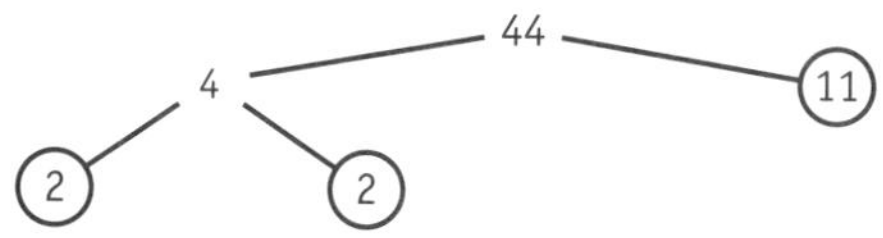

2.

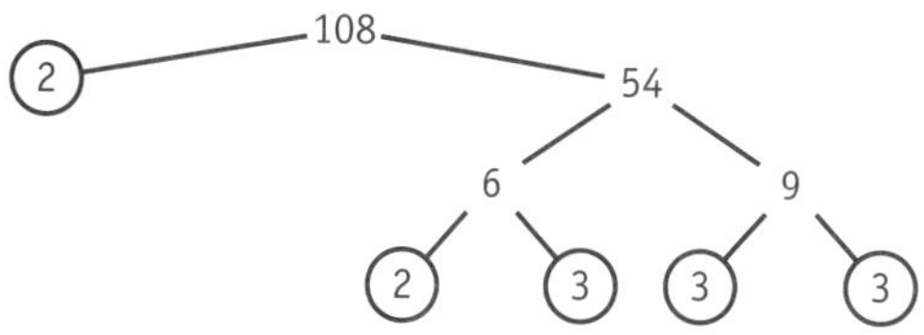

3.

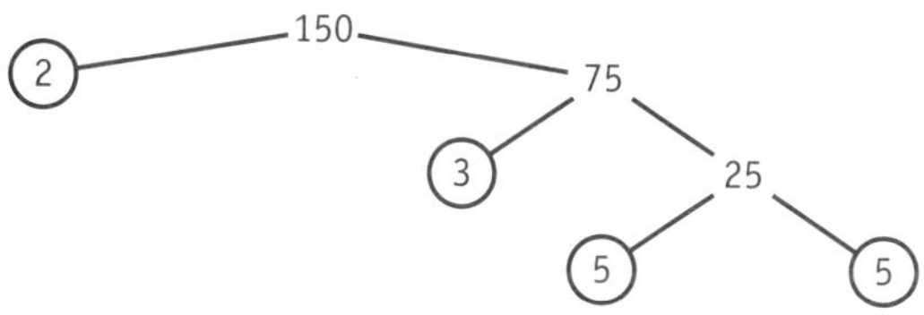

4.

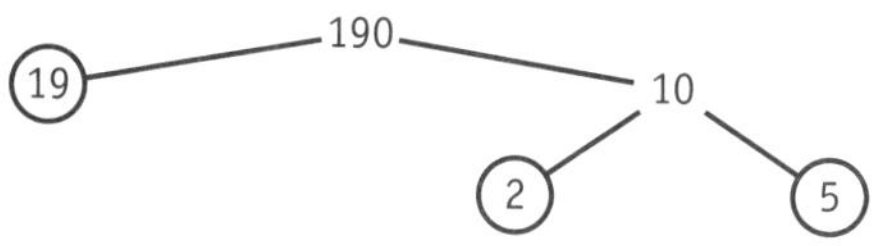

5.

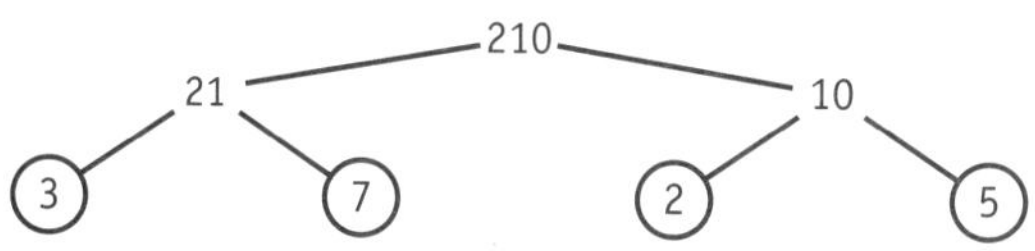

PART 3

1. The GCF is 2.
2. The GCF is 10.
3. The GCF is 1.
4. The GCF is 5.

Lesson 88—Homework

PART 1

1. the 3 rule
2. the 2, 3, and 6 rules
3. the 2, 5, and 10 rules
4. the 3 rule

PART 2

1. Area $= 8 \times 2 = 16$ sq. units
2. Area $= \frac{1}{2}(6 \times 5) = \frac{1}{2}(30) = 15$ sq. units
3. Area $= 6 \times 6 = 36$ sq. units

PART 3

Students' trees may differ depending on the facts they used, but the prime factors will be the same.

1.

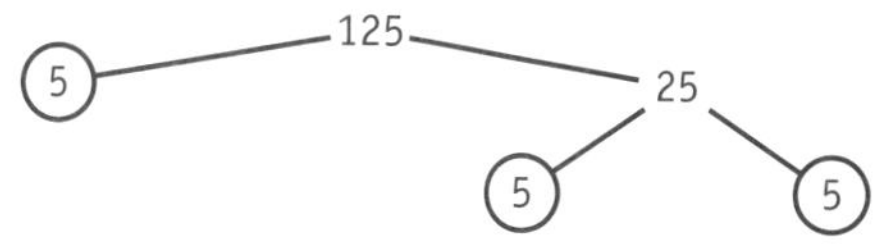

2.

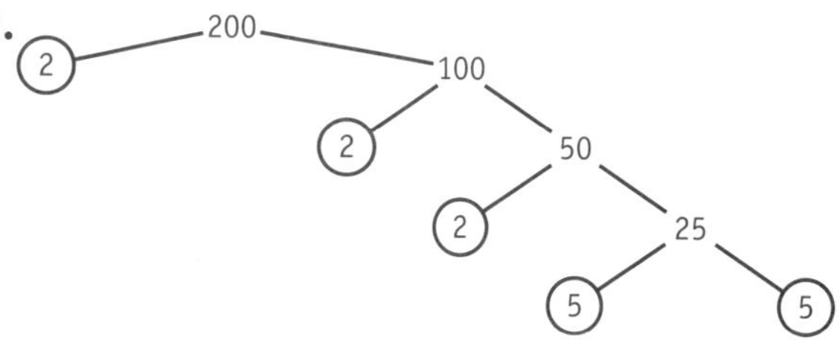

3.

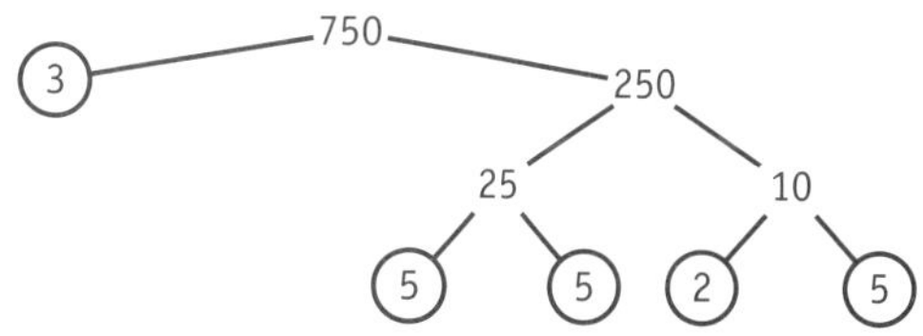

PART 4

1. The common factors are 1 and 2.
2. The common factor is 1.
3. The common factors are 1 and 3.

PART 5

1. The GCF is 2.
2. The GCF is 1.
3. The GCF is 3.

PART 6

The shapes that are congruent are C, E, and H.

PART 7

Students' trees may differ depending on the facts they used, but the prime factors will be the same.

1.

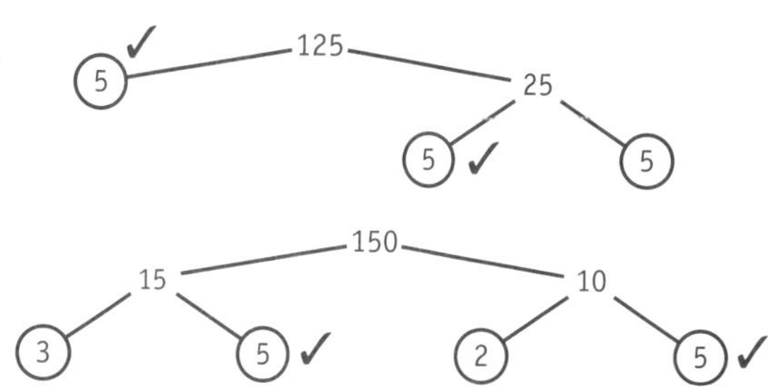

The GCF of 125 and 150 is $5 \times 5 = 25$.

2. 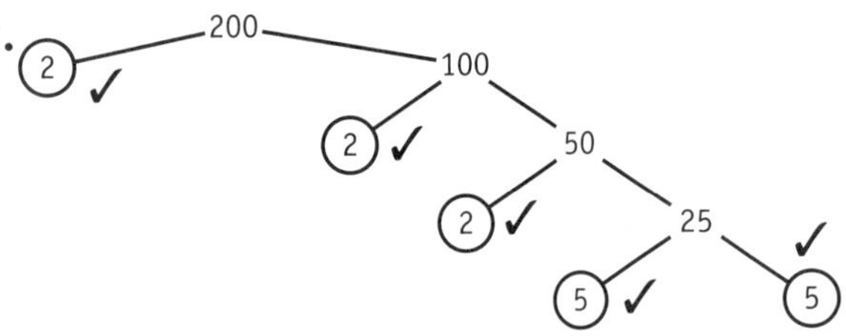

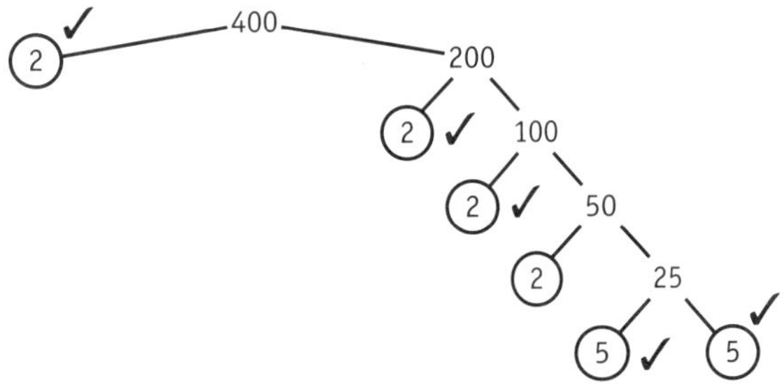

The GCF of 200 and 400 is
$2 \times 2 \times 2 \times 5 \times 5 = 200$.

3. 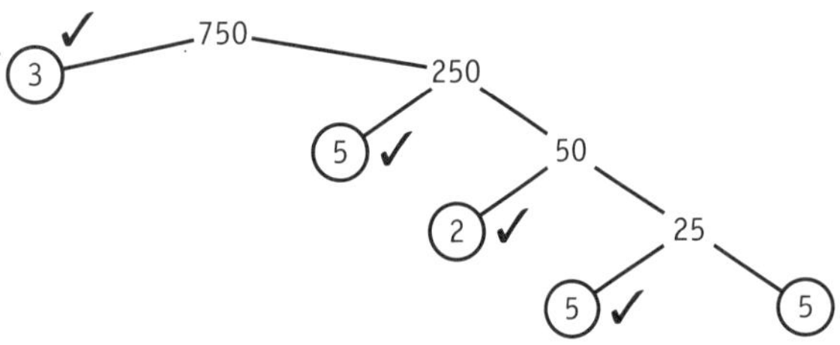

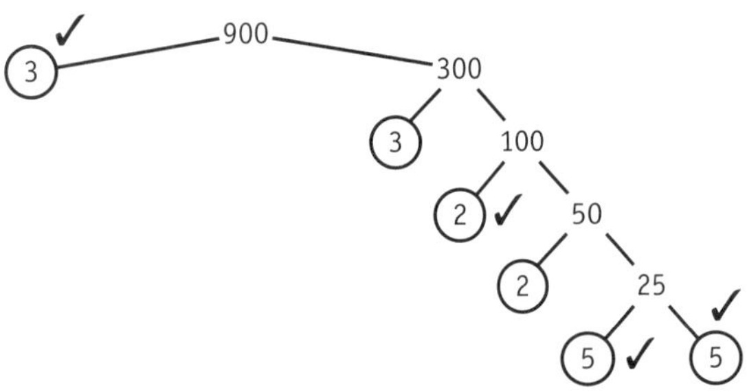

The GCF of 750 and 900 is
$3 \times 5 \times 2 \times 5 = 150$.

Lesson 89—Homework

PART 1

1. The common factors are 1 and 3.
2. The common factors are 1, 2, and 4.
3. The common factors are 1, 2, and 4.
4. The common factors are 1 and 7.

PART 2

The odd numbers are 13, 15, 103, 279, 411, 527, and 1,987.

PART 3

The even numbers are 28, 42, 116, 448, 694, and 4,289,002.

PART 4

1. The answer will be even.
2. The answer will be even.
3. The answer will be odd.
4. The answer will be odd.

PART 5

1. The answer will be even.
2. The answer will be odd.
3. The answer will be even.
4. The answer will be even.

Lesson 90—Homework

PART 1

1. 16
2. 64
3. 25
4. 36
5. 81
6. 4

PART 2

The square numbers are 4, 9, 16, 25, 36, 49, and 81.

PART 3

1. 3×3 array

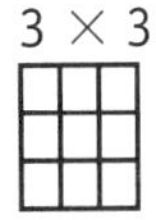

2. 4×4 array

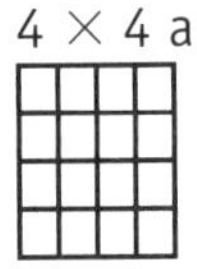

3. 5×5 array

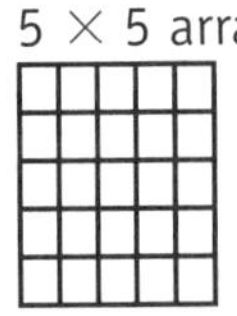

4. 2×2 array

PART 4

1. 49
2. 64
3. 16
4. 25

Lesson 91—Homework

PART 1

1. 9
2. 49
3. 1
4. 16
5. 36
6. 100

PART 2

The square numbers are 16, 25, 64, 81, 121, and 144.

PART 3

a. 1
b. 4
c. 4
d. 9
e. 9
f. 16
g. 16
h. 25
i. 25
j. 36
k. 36
l. 49

(All the answers are square numbers.)

Lesson 92—Problem-Solving

See individual student work. The area of the floor for repainting is $29 \times 15 = 435$ sq. m.

The cost is \$2 per sq. m, so the cost for this floor would be $435 \times 2 =$ \$870 for repainting.

The number of seats in the arena is 189. At a cost of \$15 per seat, the total cost would be 189×15 or \$2,835 to replace each seat.

Lesson 92—Homework

PART 1

The square numbers are 25, 49, 64, 81, 100, 121, and 144.

PART 2

1. 6

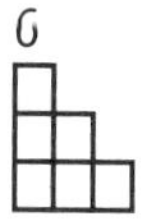

2. 10

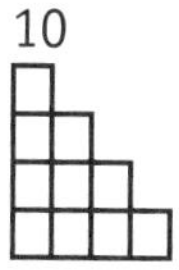

3. 15

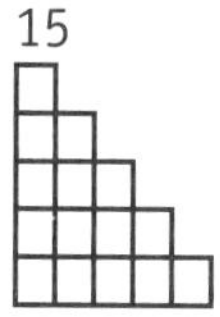

PART 3

a. 24
b. 25
c. 48
d. 49
e. 80
f. 81

Lesson 93—Homework

PART 1

1. 5
2. 4
3. 7
4. 8

PART 2

1. 10^2
2. 5^3
3. 7^2
4. 10^5

PART 3

1. 10×10
2. $2 \times 2 \times 2$
3. $3 \times 3 \times 3$
4. $10 \times 10 \times 10 \times 10 \times 10 \times 10$

PART 4

Students' trees may differ depending on the facts they used, but the prime factors will be the same.

1.

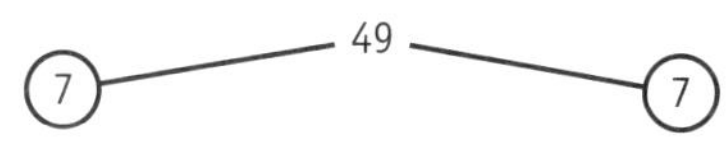

$49 = 7^2$

2.

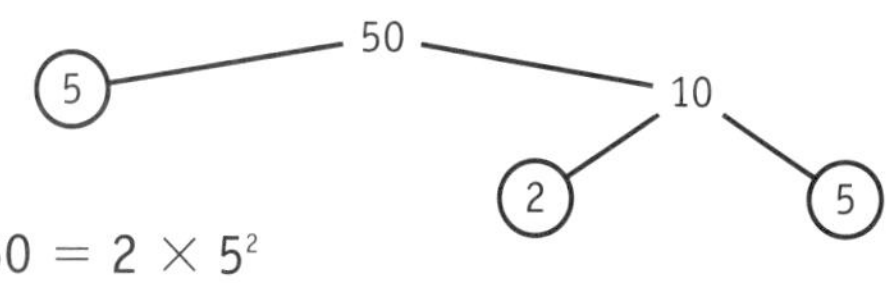

$50 = 2 \times 5^2$

3.

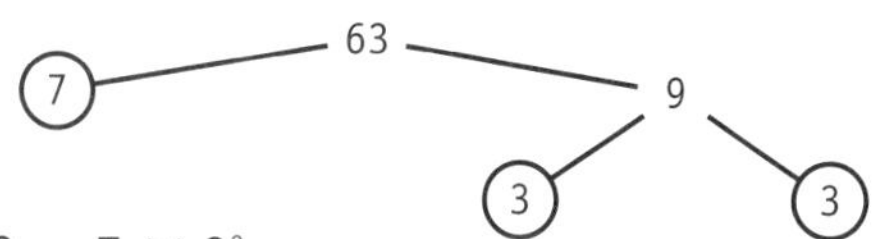

$63 = 7 \times 3^2$

4.

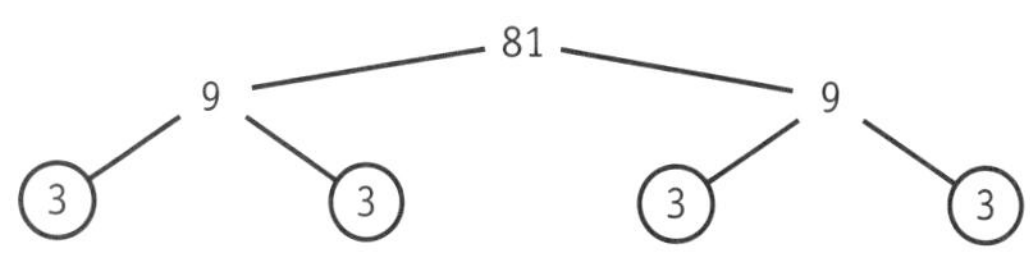

$81 = 3^4$

Lesson 94—Homework

PART 1

1. 9
2. 50
3. 70
4. 9

PART 2

1. 2^4
2. 3^6
3. 5^3
4. 10^{11}

PART 3

1. The common multiples shown are 10, 20, 30, and 40.
2. The common multiples shown are 8, 16, and 32.
3. The common multiples shown are 30 and 60.
4. The common multiples shown are 8 and 16.

PART 4
1. turn
2. turn
3. flip

Lesson 95—Homework
PART 1
1. a = 24, b = 36, c = 54
2. a = 32, b = 48, c = 72
3. a = 18, b = 36, c = 63
4. a = 40, b = 60, c = 90

PART 2
1. The LCM is 8.
2. The LCM is 63.
3. The LCM is 30.

PART 3
C shows symmetry.
A, B, and D do not.

Lesson 96—Homework
PART 1
1. a = 20, b = 30, c = 45
2. a = 24, b = 36, c = 54
3. a = 14, b = 28, c = 49
4. a = 44, b = 66, c = 99

PART 2
1. The LCM is 4.
2. The LCM is 42.
3. The LCM is 18.

Lesson 97—Homework
PART 1
1. The LCM is 24.
2. The LCM is 24.

PART 2
1. slide
2. turn
3. flip

Lesson 98—Homework
PART 1
1. The LCM is 60.
2. The LCM is 77.
3. The LCM is 36.

PART 2
1 and 2 have rotational symmetry.
3 and 4 have line symmetry.

PART 3
1. b
2. b
3. b
4. a

Lesson 99—Homework
PART 1
1. The LCM is 40.
2. The LCM is 24.

PART 2
See individual student responses. Look for discussions similar to the following:

1. Ways of describing 25
   - 5^2
   - 5×5
   - can be divided by 5
   - odd
   - square number
2. Ways of describing 100
   - 10^2
   - 10×10
   - can be divided by 5 or 10
   - even
   - square number
3. Ways of describing 237
   - can be divided by 3
   - odd
   - 79×3
   - 1×237
4. Ways of describing 558
   - can be divided by 2
   - even
   - can be divided by 3
   - can be divided by 6

Lesson 100—Homework
PART 1
1. even
2. even
3. odd
4. odd

PART 2
1. even
2. odd
3. even
4. even

PART 3
1. 25
2. 16
3. 49
4. 81

PART 4

a. 36
b. 49
c. 49
d. 64
e. 64
f. 81
g. 81
h. 100

(They are all square numbers.)

PART 5

1. 6

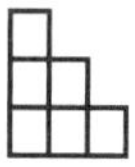

2. 10

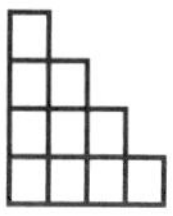

3. 15

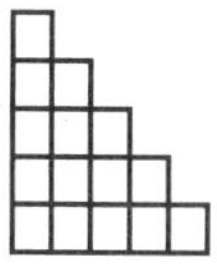

PART 6

1. 7^2
2. 3^3
3. 5^2
4. 10^5
5. 2^8

PART 7

1. 10, 20, 30, and 40
2. 8, 16, 24, and 32

PART 8

1. turn
2. flip
3. flip

PART 9

1. The LCM is 6.
2. The LCM is 12.

PART 10

1. point of rotation
2. line of symmetry
3. line of symmetry
4. point of rotation

STUDENT WORKBOOK SOLUTIONS

STUDENT WORKBOOK ANSWER KEY

UNIT 1

Addition

Lesson 1

PART A—Activity 1

1. 6,421,359

100 Millions	10 Millions	Millions	100 Thousands	10 Thousands	Thousands	100s	10s	1s
		6	4	2	1	3	5	9

2. 500,407

100 Millions	10 Millions	Millions	100 Thousands	10 Thousands	Thousands	100s	10s	1s
			5	0	0	4	0	7

3. 21,058

100 Millions	10 Millions	Millions	100 Thousands	10 Thousands	Thousands	100s	10s	1s
				2	1	0	5	8

4. 105,009,670

100 Millions	10 Millions	Millions	100 Thousands	10 Thousands	Thousands	100s	10s	1s
1	0	5	0	0	9	6	7	0

PART A—Activity 2

1. 500
2. 100,000
3. 8,000,000
4. 0
5. 70

PART A—Activity 3

How Many Digits	Write the Number
6	500,700
5	40,065
3	801
8	9,040,070
2	54

Lesson 2

PART A

1. 6,000,000
2. 3
3. 50
4. 9
5. 7
6. ten thousands place

PART B—Activity 1

1. 500 + 40 + 3
2. 600 + 0 + 5
3. 800 + 90 + 0
4. 1,000 + 0 + 0 + 5
5. 2,000 + 0 + 10 + 2
6. 5,000 + 300 + 90 + 6

PART B—Activity 2

1. 972
2. 720
3. 401
4. 5,072
5. 7,301
6. 8,013

PART C

Problem 1: The problem is asking: How many hot dogs were sold at the last two games?

There were 5,200 hot dogs sold at the ballpark last night. The attendance was 14,507. The hot dogs cost $3.75 each. The night before last, there were 4,100 hot dogs sold. The owner of the ballpark wanted to know how many hot dogs were sold at these last two games. The attendance that night was 17,129.

Problem 2: The problem is asking: How much did Sean spend at the arcade?

Sean met 5 friends at the arcade. He has $20 to spend. He spent $3 on air hockey. His friends each played 6 games. Sean spent $10 on video games. Then he spent $2 for a drink and a snack. He earned 175 tickets from the video games. His friend Kevin earned 200 tickets. Sean's mother asked him how much he spent at the arcade. What was his answer?

Lesson 3
PART A

1. 13
2. 15
3. 13
4. 15
5. 11
6. 16
7. 11
8. 16
9. 14
10. 17
11. 14
12. 17

PART B—Activity 1

1.	13	130	1,300
2.	15	150	1,500
3.	11	110	1,100
4.	16	160	1,600
5.	14	140	1,400
6.	17	170	1,700

PART B—Activity 2

Extended Fact ⟶ 80 + 50 = 130
Basic Fact ⟶ 8 + 5 = 13

PART C

Problem 1: The problem is asking: How much would Nick have saved if he had bought 25 singles instead of 25 CDs?

Nick has 25 CDs. They each cost about $15. His CD player cost $60. He likes only one song on each CD. He could buy CDs of the singles of the songs he likes for about $5 each. How much money would he have saved had he bought 25 CDs of singles instead?

Problem 2: The problem is asking: How long will it take Jared to earn 7,500 tickets at the arcade?

Jared likes to play arcade games at the mall. He collects the tickets that he earns from playing the games and saves them. He currently has 2,500 tickets. If he earns 500 tickets each time he visits the arcade, he wants to know how long it will take him to earn the big prize. Jared is saving for one of the big prizes that require 7,500 tickets. He spends about $10 each time he goes to the arcade.

Lesson 4
PART A

1.	17	170	1,700
2.	13	130	1,300
3.	13	130	1,300
4.	13	130	1,300

PART B—Activity 1

1. $\begin{array}{r} 37 \\ +\ 42 \\ \hline \end{array}$
2. $\begin{array}{r} 29 \\ +\ 95 \\ \hline \end{array}$
3. $\begin{array}{r} 58 \\ +\ 27 \\ \hline \end{array}$
4. $\begin{array}{r} 107 \\ +\ 39 \\ \hline \end{array}$

PART B—Activity 2

1. 68 + 42
2. 37 + 85
3. 59 + 11
4. 120 + 307

PART C

Problem 1: The problem is asking: How many home runs did the Blue Jays hit in June and July?

The Blue Jays hit 24 home runs in June and 28 home runs in July. In August, the team's pitcher, Emilio Sanchez, had 2 games where he struck out 15 batters. In August, the Blue Jays hit 21 home runs. The coach wants to know how many home runs the Blue Jays hit in June and July. The team is getting ready for 4 straight games at home.

Problem 2: The problem is asking: How long are the plane flights all together?

It takes 5 hours to fly from San Francisco to New York. It takes another 7 hours to fly from New York to Rome, Italy. People who travel this route usually stay in New York at a hotel by the airport and go to Rome the next day. Andrea wants to travel from San Francisco to New York to Rome. She wants to know how long the plane flights are, all together, but she does not want to spend the night in New York. She will be on a business trip, and she won't have a lot of extra time.

Lesson 5
PART A

1. 400 + 0 + 7
2. 300 + 10 + 2

3. 600 + 50 + 0
4. 1,000 + 300 + 0 + 9
5. 2,000 + 0 + 0 + 2
6. 9,000 + 0 + 90 + 6

PART B—Activity 1

1.

30	7
+ 40	2

2.

20	9
+ 90	5

3.

50	8
+ 20	7

PART B—Activity 2

1. $\begin{array}{r} 68 \\ +\ 42 \\ \hline \end{array}$
2. $\begin{array}{r} 37 \\ +\ 85 \\ \hline \end{array}$
3. $\begin{array}{r} 59 \\ +\ 11 \\ \hline \end{array}$
4. $\begin{array}{r} 165 \\ +\ 88 \\ \hline \end{array}$
5. $\begin{array}{r} 307 \\ +\ 500 \\ \hline \end{array}$

PART C

1. The problem is asking: How much did the boys spend at the movies?

   Leroy and Jamaal went to the movies on Saturday afternoon. Each ticket cost $5.50, and they bought snacks. Sarah and Julia were at the movies that afternoon, too. Leroy bought popcorn and a drink for $5, and Jamaal just bought a drink for $3. When Leroy got home, his mom wanted to know how much he and Jamaal spent at the movies. Leroy's mom had just gotten home from work.

2. The problem is asking: How many days did it take the climbers to hike to the top of the mountain?

   Climbing a tall mountain can take several days and a lot of planning. A group of climbers from Kenya wanted to climb one of the tallest mountains in the world. They hiked up the mountain two days before they reached the first base camp. It took them another two days to reach the second base camp. It was very cold that morning. Many of the climbers took out jackets they didn't think they would need. Finally, they hiked that day to the top of the mountain. How many days did it take them to hike to the top of the mountain?

Lesson 6

PART A

1. 130
2. 15
3. 1,300
4. 13
5. 11
6. 150
7. 1,500
8. 1,700
9. 110
10. 17
11. 170
12. 1,100

PART B

1.

30	7
+ 40	2
70	9

79

2.

60	1
+ 20	5
80	6

86

3.

70	4
+ 10	3
80	7

87

Lesson 7

PART A

1. 13
2. 13
3. 15
4. 15
5. 110
6. 110
7. 16
8. 16
9. 14
10. 14
11. 170
12. 170

PART B—Activity 1

1.

300	30	7
+ 500	40	2
800	70	9

879

2.

400	60	1
+ 200	20	5
600	80	6

686

3.

700	80	0
+ 100	0	9
800	80	9

889

PART B—Activity 2

1. The number is 245.
2. The number is 524.
3. The number is 401.

Lesson 8

PART A

1. 900
2. 90
3. 1,000
4. 170
5. 110
6. 1,100
7. 80
8. 800
9. 900
10. 90
11. 1,700
12. 170

PART B—Activity 1

1.

300	10	2
+ 400	50	7
700	60	9

769

2.

200	0	7
+ 600	90	1
800	90	8

898

3.

500	60	6
+ 300	0	3
800	60	9

869

PART B—Activity 2

1. 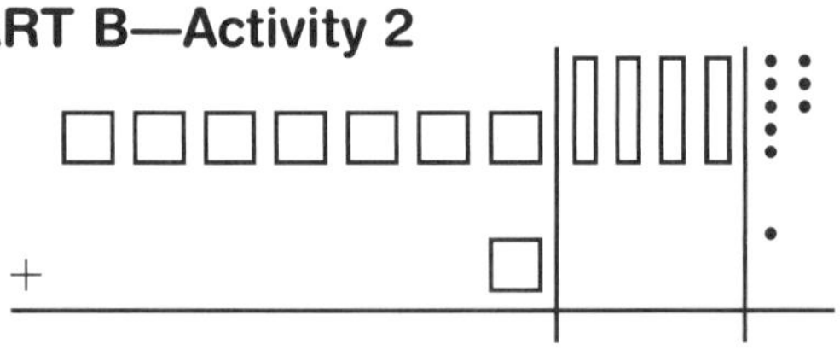

2. 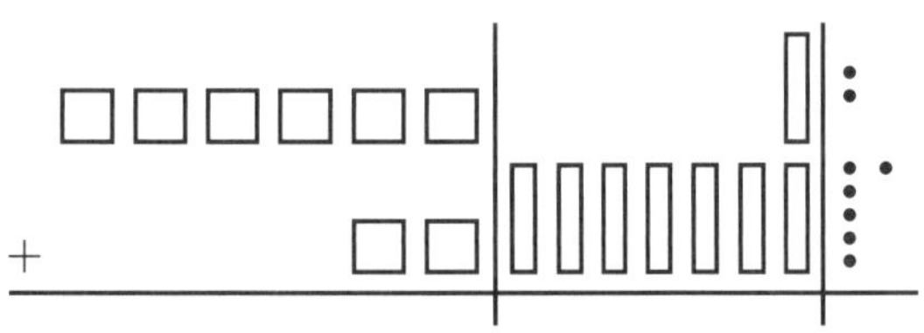

3. 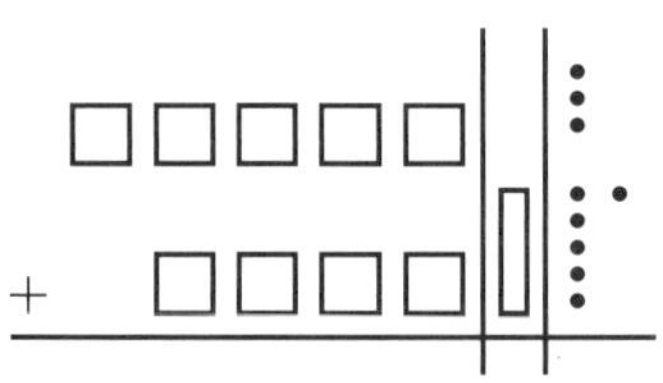

PART B—Activity 3

1. 646 + 300 = 946

$$\begin{array}{r} 646 \\ +\ 300 \\ \hline 946 \end{array}$$

2.

$$\begin{array}{r} 303 \\ +\ 212 \\ \hline 515 \end{array}$$

3.

$$\begin{array}{r} 326 \\ +\ 412 \\ \hline 738 \end{array}$$

Lesson 9

PART A

1. 8 80 800
2. 9 90 900
3. 9 90 900
4. 4 40 400

PART B

1.

10 (carried)	
50	7
+ 20	8
	15
80	5

85

2.

10 (carried)	
40	5
+ 10	6
	11
60	1

61

PART C

1. The problem is asking: How much more money did the top movie make compared to the number 2 movie?

   This list shows the top 5 movies and how much they made over 1 weekend at the end of November, 2001.

Movies	Weekend Sales
Harry Potter	$82 million
Monster, Inc.	$32 million
Spy Game	$30 million
Black Knight	$15 million
Shallow Hal	$12 million

One movie company was interested in how much more money the top movie made than the number 2 movie. This company was thinking about making a movie similar to *Harry Potter*. The sales amounts listed in the chart are for just 1 weekend. The sales amounts do not include the total money the movies have made since they were released.

2. The problem is asking: How much money did the movie make since it came out?

   The movie that was tenth that weekend was *Life as a House*. It made $3 million dollars over the weekend. The movie had been out for 5 weeks, and before that weekend it had made a total of $9 million. The producer wanted to know how much money the movie had made since it first came out, including the $3 million weekend. She was planning to use this number to figure out when the movie would make a profit.

Lesson 10

PART A

1. 11 110 1,100
2. 10 100 1,000
3. 5 50 500
4. 8 80 800

PART B—Activity 1

1.
```
  10
  40 | 7
+ 30 | 5
     | 12
  80 | 2
82
```

2.
```
  10
  20 | 9
+ 10 | 8
     | 17
  40 | 7
47
```

PART B—Activity 2

1.
```
   1
   47
 + 35
   82
```

2.
```
   1
   29
 + 18
   47
```

PART B—Activity 3

Expanded Addition vs. Traditional Addition

Similarities:

- Both involve regrouping.
- Both involve "carrying."
- Both result in the same answer.

Differences:

- One requires less writing than the other.
- One shows a carry of "10"; the other shows a tally mark for the carry.

PART C

1.
```
   5,976
 +   398
   6,374
```

2.
```
   62,495
 + 28,786
   91,281
```

3.
```
   1,323,597
 +   565,318
   1,888,915
```

Lesson 11

PART A

1. 6
2. 6
3. 12
4. 12
5. 7
6. 7
7. 4
8. 8
9. 9
10. 8
11. 18
12. 18

PART B—Activity 1

Expanded Form:

1.
```
  10
  30 | 9
+ 40 | 8
     | 17
  80 | 7
87
```

2.
```
  10
  20 | 4
+ 50 | 8
     | 12
  80 | 2
82
```

3.
```
  10
  70 | 6
+ 10 | 7
     | 13
  90 | 3
93
```

Traditional Method:

1.
```
   1
   39
 + 48
   87
```

2.
```
   1
   24
 + 58
   82
```

3.
```
   1
   76
 + 17
   93
```

PART B—Activity 2

1.
```
   554
 +  87
   641
```

2.
```
   2,987
 + 6,983
   9,970
```

3.
```
   13,784
 +  7,228
   21,012
```

PART C

Check individual student work for valid questions and answers.

Sample Question: How much more money did KRAA make on advertising than KWWQ?

Answer: \$4 million – \$2 million = \$2 million. KRAA made \$2 million more than KWWQ on advertising.

Lesson 12

PART A

1. 13
2. 130
3. 1,300
4. 7
5. 70
6. 700
7. 4
8. 40
9. 400
10. 5
11. 50
12. 500

PART B

Expanded Form:

1.

	10	
200	30	9
+	20	5
		14
200	60	4

264

2.

100	10	
500	70	6
+	40	7
		13
	120	3
600	20	3

623

3.

100	10	
100	70	9
+ 700	20	4
		13
	100	3
900	0	3

903

Traditional Method:

1.
$$\begin{array}{r} {}^{1}\\ 239 \\ +\ 25 \\ \hline 264 \end{array}$$

2.
$$\begin{array}{r} {}^{1\,1}\\ 576 \\ +\ 47 \\ \hline 623 \end{array}$$

3.
$$\begin{array}{r} {}^{1\,1}\\ 179 \\ +\ 724 \\ \hline 903 \end{array}$$

PART C

Check individual student work for valid questions and answers.

1. Sample Question: What is the difference in sales between Week 6 and Week 4?
   Answer:
   Week 6 ⟶ \$600
   Week 4 ⟶ \$250
   600 − 250 = \$350
   Week 6 sales were bigger than Week 4 sales by \$350.
2. Sample Question: If the trend continues, when do you think the company will stop advertising?
   Answer:
   Because the sales have been dropping by about \$50 per week, and at Week 10 the sales are at \$350, the company will stop advertising after Week 13.
   Week 11 ⟶ 350 − 50 = \$300
   Week 12 ⟶ 300 − 50 = \$250
   Week 13 ⟶ 250 − 50 = \$200
   Week 14 ⟶ 200 − 50 = \$150

Lesson 13

PART A

1. 1,600
2. 120
3. 80
4. 800
5. 700
6. 70
7. 700
8. 1,600
9. 800
10. 120
11. 1,200
12. 120

PART B

1. 49 + 82 = ____
   - Round 49 to 50.
   - Round 82 to 80.
   - The extended fact is 50 + 80 = 130.
   - Calculator Answer: 49 + 82 = 131
2. 92 + 69 = ____
   - Round 92 to 90.
   - Round 69 to 70.
   - The extended fact is 90 + 70 = 160.
   - Calculator Answer: 92 + 69 = 161
3. 697 + 921 = ____
   - Round 697 to 700.
   - Round 921 to 900.
   - The extended fact is 700 + 900 = 1,600.
   - Calculator Answer: 697 + 921 = 1,618
4. 401 + 837 = ____
   - Round 401 to 400
   - Round 837 to 800
   - The extended fact is 400 + 80 = 1200.
   - Calculator Answer: 401 + 837 = 1,238

PART C

Expanded Form:

1.

100	10	
100	70	6
+	20	7
		13
	100	3
200	0	3

203

2.

	10	
500	70	8
+	10	3
		11
500	90	1

591

3.

	10	
100	40	9
+ 700	20	5
		14
800	70	4

874

4.

100	10	
200	0	8
+ 300	90	4
		12
	100	2
600	0	2

602

Traditional Method:

1. $\begin{array}{r} {}^{1\,1}\\ 176 \\ +\ 27 \\ \hline 203 \end{array}$

2. $\begin{array}{r} {}^{1}\ \\ 578 \\ +\ 13 \\ \hline 591 \end{array}$

3. $\begin{array}{r} {}^{1}\ \\ 149 \\ +\ 725 \\ \hline 874 \end{array}$

4. $\begin{array}{r} {}^{1\,1}\\ 208 \\ +\ 394 \\ \hline 602 \end{array}$

PART D

Problem 1: Extended fact equation ⟶
600 + 600 = 1,200
Actual Equation ⟶ 577 + 621 = 1,198
What is the difference? Only 2

Why is the difference so small? We rounded one number up by about 30 and one number down by about 30.

Problem 2: Extended fact equation ⟶
100 + 100 = 200
Actual Equation ⟶ 142 + 135 = 277
What is the difference? 77

Why is the difference so large? We rounded one number down by about 30 and one number down by about 40 so the difference is around 70.

Lesson 14

PART A

1. 30 + 30 = 60
2. 50 + 10 = 60
3. 400 + 300 = 700
4. 600 + 800 = 1,400

PART B

The first problem, 29 + 87, is incorrect. The student forgot to add the carry in the tens place. The correct answer is 116.

Lesson 15

PART A

1. 6
2. 6
3. 60
4. 600
5. 5
6. 5
7. 500
8. 50
9. 9
10. 9
11. 900
12. 90

PART B (Check individual student work for correct expanded form.)

1. 63
2. 235
3. 1,333

PART C—Activity 1

1. 79 + 92 = ____
   - Round 79 to 80.
   - Round 92 to 90.
   - The extended fact is 80 + 90 = 170.
   - Calculator Answer: 79 + 92 = 171
2. 642 + 721 = ____
   - Round 642 to 600.
   - Round 721 to 700.
   - The extended fact is 600 + 700 + 1,300.
   - Calculator Answer: 642 + 721 = 1,363

PART C—Activity 2

Problem 1: Extended fact equation ⟶
500 + 600 = 1,100
Actual Equation ⟶ 498 + 631 = 1,129
The answers differ by 29.

Problem 2: Extended fact equation ⟶
1,900 + 100 = 2,000
Actual Equation ⟶ 1,906 + 95 = 2,001
The answers differ by 5.

PART D

The first two problems are incorrect.

In the first problem, the student forgot to add the carry in the tens place. The correct answer is 127.

In the second problem, the student added an additional hundred. There was actually no carry into the hundreds place. The correct answer is 352.

Lesson 16

PART A

1. 50
2. 50
3. 130
4. 130
5. 70
6. 70
7. 50
8. 50
9. 80
10. 80
11. 180
12. 180

PART B—Activity 1

Step 1: 30 + 7 + 40 + 2
Step 2: 30 + 40 + 7 + 2
Step 3: 70 + 9
Answer: 79

PART B—Activity 2

55 + 22 = 50 + 5 + 20 + 2
50 + 20 + 5 + 2
70 + 7
77

34 + 44 = 30 + 4 + 40 + 4
30 + 40 + 4 + 4
70 + 8
78

63 + 25 = 60 + 3 + 20 + 5
60 + 20 + 3 + 5
80 + 8
88

31 + 38 = 30 + 1 + 30 + 8
30 + 30 + 1 + 8
60 + 9
69

Lesson 17

PART A—Activity 1

1. 130
2. 1,200
3. 14
4. 13
5. 14
6. 12
7. 130
8. 130
9. 12
10. 1,600
11. 1,400
12. 12

PART A—Activity 2

1. 49
2. 9
3. 67
4. 160
5. 13
6. 57
7. 130
8. 49
9. 84
10. 87
11. 110
12. 78

PART B—Activity 1

1. 30 + 7 + 40 + 2
   30 + 40 + 7 + 2
   70 + 9
   79
2. 60 + 1 + 20 + 5
   60 + 20 + 1 + 5
   80 + 6
   86

PART B—Activity 2 (Check individual student work for correct place value block notation.)

1. 33
2. 83
3. 64
4. 95

Lesson 18

PART A

1. 1,400
2. 140
3. 1,100
4. 110
5. 120
6. 1,200
7. 130
8. 1,300
9. 1,200
10. 120
11. 1,700
12. 170

PART B—Activity 1

1. 400 + 30 + 7 + 40 + 2
   400 + 30 + 40 + 7 + 2
   400 + 70 + 9
   479
2. 100 + 20 + 3 + 400 + 10 + 0
   100 + 400 + 20 + 10 + 3 + 0
   500 + 30 + 3
   533

PART B—Activity 2 (Check individual student work for correct place value block notation.)

1. 779
2. 818
3. 946
4. 591

Lesson 19

PART A

1. 2 20 200
2. 6 60 600
3. 8 80 800
4. 11 110 1,100

PART B

1. 50 + 7 + 20 + 1
   50 + 20 + 7 + 1
   70 + 8
   78
2. 30 + 2 + 40 + 3
   30 + 40 + 2 + 3
   70 + 5
   75
3. 100 + 20 + 3 + 40 + 5
   100 + 20 + 40 + 3 + 5
   100 + 60 + 8
   168
4. 200 + 30 + 4 + 50 + 1
   200 + 30 + 50 + 4 + 1
   200 + 80 + 5
   285
5. 100 + 10 + 6 + 300 + 20 + 1
   100 + 300 + 10 + 20 + 6 + 1
   400 + 30 + 7
   437
6. 200 + 0 + 7 + 400 + 40 + 2
   200 + 400 + 0 + 40 + 7 + 2
   600 + 40 + 9
   649

Lesson 20

PART A

1.	6	60	600	**3.**	5	50	500
2.	8	80	800	**4.**	13	130	1,300

PART B

1. 500
2. 10
3. 6,000
4. 3

PART C (Note that the zero place holders are optional.)

1. 200 + 90 + 3
2. 300 + 0 + 7
3. 500 + 90 + 8
4. 100 + 10 + 0
5. 500 + 60 + 0
6. 1,000 + 0 + 90 + 8

PART D

1.	110	11	1100
2.	900	90	9
3.	8	80	800

PART E (Check individual student work for correct expanded form.)

1. 73
2. 598
3. 539
4. 666

PART F

1. 20 + 8 + 60 + 1
   20 + 60 + 8 + 1
   80 + 9
   89
2. 30 + 7 + 20 + 2
   30 + 20 + 7 + 2
   50 + 9
   59
3. 100 + 50 + 2 + 30 + 5
   100 + 50 + 30 + 2 + 5
   100 + 80 + 7
   187

PART G

1. Estimate: 10 + 60 = 70
   Exact Answer: 11 + 59 = 70
2. Estimate: 200 + 20 = 220
   Exact Answer: 199 + 24 = 223
3. Estimate: 400 + 400 = 800
   Exact Answer: 387 + 421 = 808
4. Estimate: 500 + 200 = 700
   Exact Answer: 501 + 211 = 712

PART H

Problems 2 and 3 are incorrect.

In Problem 2, the student added one too many tens in the tens place. Instead of adding the carry of 1 ten, the student added the carry of 2 tens. The answer should be 106.

In Problem 3, the student forgot to add the carry in the tens place. The correct answer is 283.

UNIT 2

STUDENT WORKBOOK ANSWER KEY

Subtraction

Lesson 21

PART A

1. 9 90 900
2. 9 90 900
3. 6 60 60
4. 6 60 600

PART B

1. 7 + 8 = 15
8 + 7 = 15
15 − 7 = 8
15 − 8 = 7

2. 2 + 9 = 11
9 + 2 = 11
11 − 2 = 9
11 − 9 = 2

3. 6 + 7 = 13
7 + 6 = 13
13 − 7 = 6
13 − 6 = 7

4. 9 + 5 = 14
5 + 9 = 14
14 − 9 = 5
14 − 5 = 9

PART C

1.

	60	9
−	20	7
	40	2

42

2.

	90	7
−	70	5
	20	2

22

3.

	70	9
−	10	6
	60	3

63

PART D

Part D is a game. You may want to assess student knowledge of extended facts by holding a class discussion about the outcome of individual student games.

Lesson 22

PART A

1. 15 15 15 15
2. 15 15 15 15
3. 7 7 7 7

PART B

1.

	60	9
−	20	7
	40	2

42

2.

	50	7
−	30	6
	20	1

21

3.

	90	6
−	30	2
	60	4

64

PART C

1. 70 + 80 = 150
80 + 70 = 150
150 − 70 = 80
150 − 80 = 70

2. 200 + 900 = 1,100
900 + 200 = 1,100
1,100 − 200 = 900
1,100 − 900 = 200

3. 60 + 90 = 150
90 + 60 = 150
150 − 60 = 90
150 − 90 = 60

4. 300 + 900 = 1,200
900 + 300 = 1,200
1,200 − 300 = 900
1,200 − 900 = 300

Lesson 23

PART A

1. 60 60 600
2. 9 90 900
3. 11 110 1,100
4. 6 60 600

PART B

1.

	400	20	8
−	200	10	6
	200	10	2

212

2.

	300	60	7
−	100	50	3
	200	10	4

214

3.

500	80	9
− 300	70	2
200	10	7

217

PART C (See student work for correct use of number lines.)

We round 56 to 60.
We round 17 to 20.
The extended fact equation is 60 − 20 = 40.
The exact answer is 39.

The estimate is very close compared to the exact answer.

The estimate is larger because we rounded the first number up by four and the second number up by three, creating a larger difference.

Lesson 24

PART A

1. 11 11 11 11
2. 110 110 110 110
3. 1,100 1,100 1,100 1,100

PART B

1.

60	4
− 10	6

→

50 + 10	4
− 10	6

50	14
− 10	6
40	8

→ 48

2.

70	5
− 50	9

→

60 + 10	5
− 50	9

60	15
− 50	9
10	6

→ 16

3.

80	2
− 70	7

→

70 + 10	2
− 70	7

70	12
− 70	7
	5

→ 5

PART C—Activity 1

1. 

2. 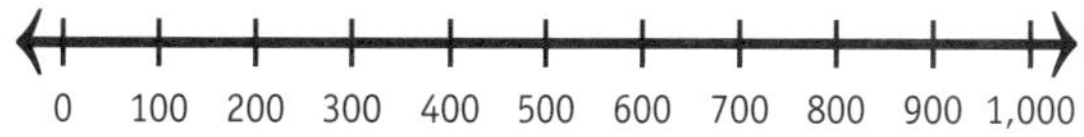

3. 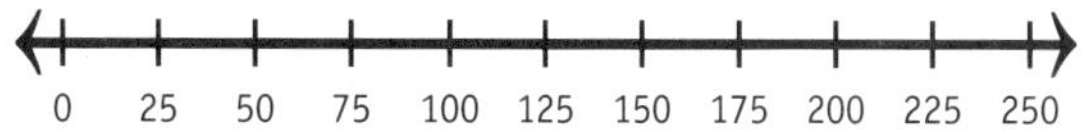

PART C—Activity 2

1. 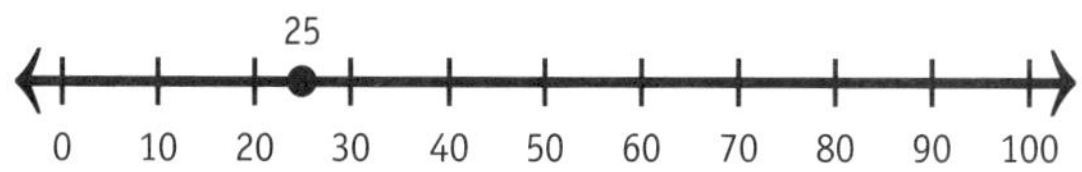

2. 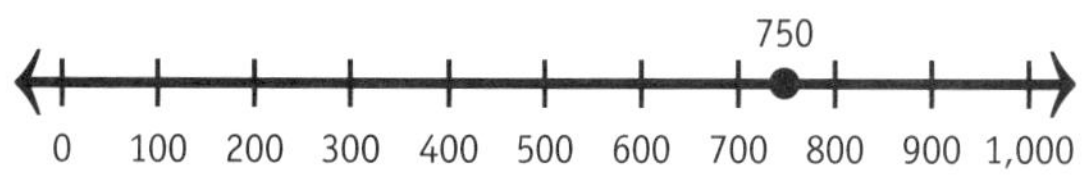

3. 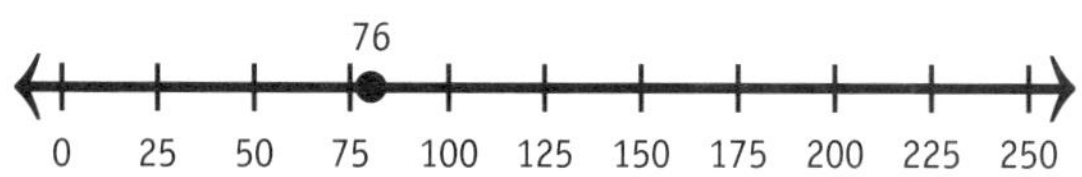

4. 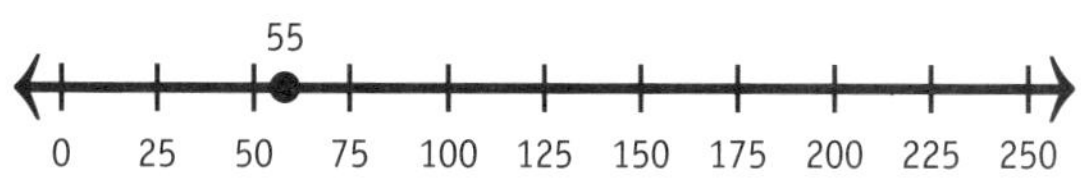

Lesson 25

PART A

1. 170 170 170 170
2. 1,500 1,500 150 150
3. 7,000 7,000 7,000 7,000

PART B

1.

50	2
− 10	8

→

40 + 10	2
− 10	8

40	12
− 10	8
30	4

→ 34

2.

90	1
− 40	3

→

80 + 10	1
− 40	3

80	11
− 40	3
40	8

→ 48

PART C

1.

```
   4 1
  4 5̸ 2
− 1 2 7
  3 2 5
```

2.

```
 4 1
 5̸ 6 4
−  8 3
 4 8 1
```

PART D

1.

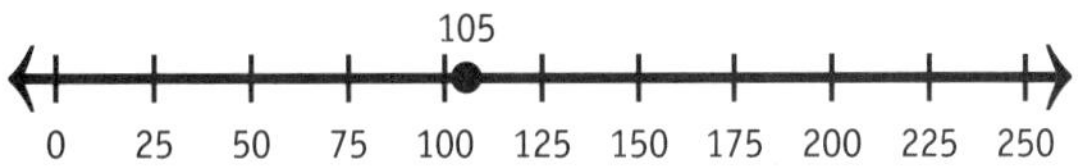

2.

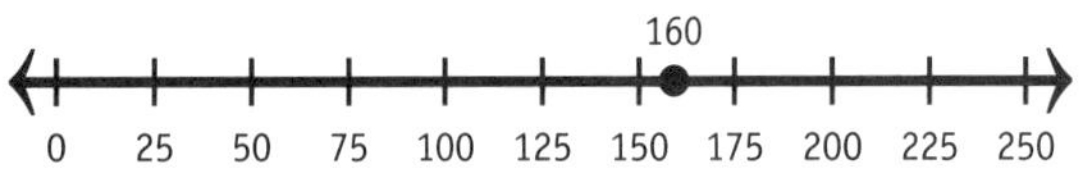

3.

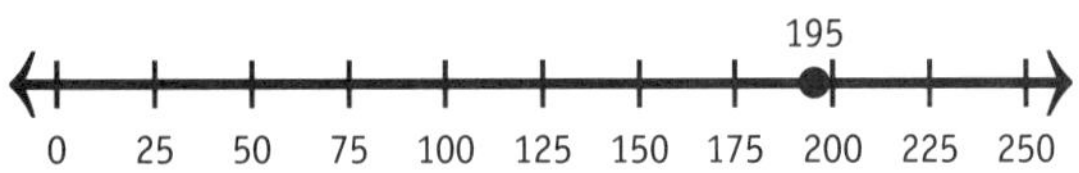

4.

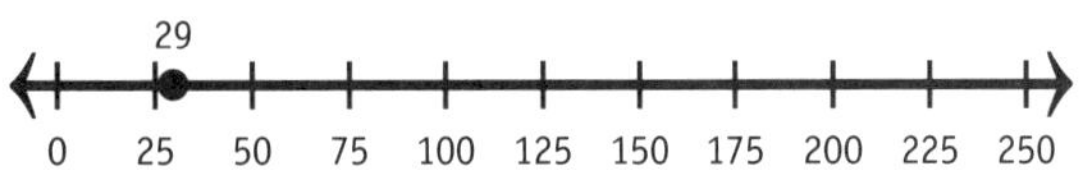

5.

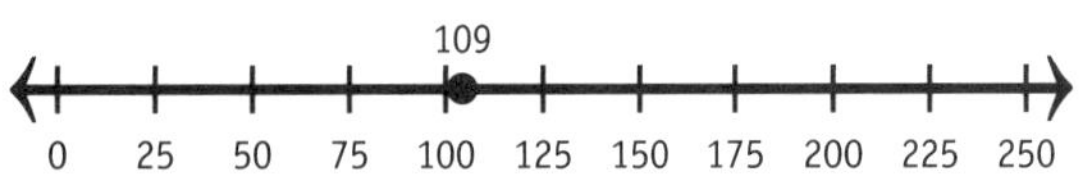

Lesson 26

PART A

1.	160	160	160	160
2.	1,400	1,400	140	140
3.	7,000	7,000	7,000	7,000

PART B

1.

300	50	2
− 100	80	1

200 + 100	50	2
− 100	80	1

200	100 + 50	2
− 100	80	1

200	150	2
− 100	80	1
100	70	1

171

2.

900	10	5
− 700	80	3

800 + 100	10	5
− 700	80	3

800	100 + 10	5
− 700	80	3

800	110	5
− 700	80	3
100	30	2

132

PART C

1. Front-End
   Estimate: 80 − 20 = 60
   Actual Answer: 57
2. Front-End
   Estimate: 300 − 100 = 200
   Actual Answer: 199
3. Quarter
   Estimate: 375 − 150 = 225
   Actual Answer: 225
4. Quarter
   Estimate: 600 − 325 = 275
   Actual Answer: 272
5. Quarter
   Estimate: 975 − 725 = 250
   Actual Answer: 245

Lesson 27

PART A

1. 50
2. 75
3. 200
4. 50
5. 100
6. 50
7. 25
8. 125
9. 50

PART B —Activity 1

```
 4 1
 5̸ 2 7
− 2 3 5
 2 9 2
```

PART B—Activity 2

500	20	7
− 200	30	5

400 + 100	20	7
− 200	30	5

400	100 + 20	7
− 200	30	5

400	120	7
− 200	30	5
200	90	2

292

Sample responses to "How do the methods differ?"

1. The expanded method involves a lot more steps.
2. The expanded method breaks the numbers into place values.
3. The expanded method shows the borrowing step more clearly.
4. The traditional method uses shorthand to show the borrowing step.
5. The expanded method is easier to follow but harder to do by yourself.
6. The expanded method does not lead to as many mistakes as the traditional method.

Lesson 28

PART A

1.	16	16	16	16
2.	160	160	160	160
3.	100	100	100	100

PART B

1.

20	8
− 10	6
10	2

12

2.

300	60	7
−	50	3
300	10	4

314

3.

400	80	9
− 300	50	3
100	30	6

136

PART C

1. Quarter
   Estimate: 575 − 325 = 250
   Exact Answer: 246
2. Front-End or Quarter
   Estimate: 400 − 200 = 200
   Exact Answer: 198
3. Quarter
   Estimate: 450 − 225 = 225
   Exact Answer: 227
4. Front-End
   Estimate: 900 − 700 = 200
   Exact Answer: 192

PART D

1. 25
2. 681
3. 425
4. 673

Lesson 29

PART A

1.	12	12	12	12
2.	120	120	120	120
3.	13	13	13	13
4.	130	130	130	130

PART B—Activity 1

1.

300	50	11
− 100	10	7
200	30	4

234

2.

400	00	4
−	20	7
300	70	7

377

PART B—Activity 2

1. The answer should be 234. The student reversed the digits in the ones place. The students subtracted 7 − 1 = 6 instead of realizing that it was 1 − 7 and the he or she needed to borrow.
2. The correct answer should be 377. The student subtracted correctly in the ones place, borrowing a ten from the tens place, 14 − 7 = 7. But then the student added the remaining numbers instead of subtracting.

Lesson 30
PART A

1.	13	13	13	13
2.	130	130	130	130
3.	11	11	11	11
4.	110	110	110	110

PART B—Activity 1

1.

800	90	6
− 200	40	8
600	40	8

648

2.

600	00	1
− 200	40	0
300	60	1

361

PART B—Activity 2

1. The answer should be 648. The student forgot to adjust the tens place after borrowing. There was a 9 in the tens place, and after the borrowing, it should have been an 8. But the student just left it as a 9 and subtracted 9 − 4 = 5. Notice that the incorrect answer is EXACTLY 10 more than the correct answer. This clearly shows that the tens place was not adjusted after borrowing.
2. The answer should be 361. This time the student adjusted the tens place without needing to. There was no borrowing necessary in the ones place: 1 − 0 = 1. For the tens place the student needed to borrow from the hundreds place, and the next subtraction should have been 10 − 4 = 6. This is where the student went wrong. The student wrote 9 − 4 = 5. The hundreds place needed to be adjusted, which it was, and the final subtraction was 5 − 2 = 3. Notice that the incorrect answer is EXACTLY 10 less than the correct answer. This clearly shows that the tens place was adjusted for borrowing when it was not supposed to be.

Lesson 31
PART A

1. 25 + 75 = 100
   75 + 25 = 100
   100 − 25 = 75
   100 − 75 = 25
2. 125 + 50 = 175
   50 + 125 = 175
   175 − 50 = 125
   175 − 125 = 50

PART B

1. Estimation: 300 − 150 = 150
   Exact Answer: 159
2. Estimation: 8,000 − 6,000 = 2,000
   Exact Answer; 1,768
3. Estimation: 6,000 − 1,000 = 5,000
   Exact Answer: 4,959
4. Estimation: 10,000 − 1,000 = 9,000
   Exact Answer: 8,901
5. Estimation: 5,100 − 100 = 5,000
   Exact Answer: 5,012

Lesson 32
PART A

1.	15	15	15	15
2.	150	150	150	150
3.	11	11	11	11
4.	1,100	1,100	1,100	1,100

PART B (See individual student work for correct use of the number lines.)

1. Round 3,056 to 3,000
   Round 1,917 to 2,000
   Extended Fact Equation:
   3,000 − 2,000 = 1,000
   Exact Answer: 1,139
2. Round 4,001 to 4,000
   Round 3,078 to 3,000
   Extended Fact Equation:
   4,000 − 3,000 = 1,000
   Exact Answer: 923
3. Round 10,056 to 10,000
   Round 4,895 to 5,000
   Extended Fact Equation:
   10,000 − 5,000 = 5,000
   Exact Answer: 5,161

Lesson 33
PART A

1.	1	1	1
2.	5	95	995
3.	9	99	999
4.	5	5	5

PART B (See individual student work for correct use of number lines.)

1. 201
2. 198
3. 205
4. 195

Lesson 34

PART A

1. 25 + 75 = 100
   75 + 25 = 100
   100 − 75 = 25
   100 − 25 = 75
2. 250 + 500 = 750
   500 + 250 = 750
   750 − 250 = 500
   750 − 500 = 250

PART B

1. Method: Estimation and Calculator
   Estimate: 5,000 − 2,000 = 3,000
   Exact Answer: 3,009
2. Method: Mental Math
   You think: 2 more is 200 and 300 more is 500. 2 + 300 = 302.
3. Method: Estimation and Calculator
   Estimate: 250 − 150 = 100
   Exact Answer: 112
4. Method: Estimation and Calculator
   Estimate: 1,200 − 1,000 = 200
   Exact Answer: 226
5. Method: Mental Math
   You think: 1 more is 2,000, 3,000 more is 5,000, and 1 more is 5,001.
   1 + 3,000 + 1 = 3,002

Lesson 35

PART A

1.	75	75	75	75
2.	750	750	750	750
3.	100	100	100	100

PART B

1.

	60	7
−	30	2
	30	5

35

2.

800	50	6
−	40	5
800	10	1

811

3.

700	70	6
− 200	20	3
500	50	3

553

PART C

1. Quarter
   250 − 175 = 75
   Actual Answer: 72
2. Front-End
   400 − 300 = 100
   Actual Answer: 102
3. Quarter
   325 − 200 = 125
   Actual Answer: 128
4. Quarter
   750 − 525 = 225
   Actual Answer: 227

PART D (See individual student work to check regrouping.)

1. 59
2. 371
3. 392
4. 606

PART E

1. Correct
2. Incorrect. The correct answer is 371. The student reversed the digits and subtracted 9 − 6 = 3. The student should have subtracted 6 - 9, which requires regrouping.
3. Correct
4. Incorrect. The correct answer is 606. The student borrowed from the hundreds place when it was not necessary to borrow. Notice that the answer is 100 more than the incorrect answer. That shows clearly that the student took away 100 too many.

PART F

1. Method: Estimation and Calculator
   Estimate: 5,000 − 2,000 = 3,000
   Exact Answer: 3,009
2. Method: Good Number Sense and Number Line
   You think: 5 more makes 400 and 200 more makes 600. 5 + 200 = 205.
3. Method: Estimation and Calculator.
   Estimate: 875 − 525 = 350
   Exact Answer: 348

UNIT 3

STUDENT WORKBOOK ANSWER KEY

Multiplication

Lesson 36

Part A

1. 27
2. 72
3. 7
4. 27
5. 9
6. 9
7. 72
8. 6
9. 4

PART B (See individual student work for correct use of number lines.)

1. 9
2. 10
3. 8

PART C—Activity 1

1.	18	180	1,800
2.	27	270	2,700
3.	56	560	5,600
4.	24	240	2,400
5.	40	400	4,000

PART C—Activity 2

1.	8	80	800
2.	8	80	800
3.	9	9	9

PART C—Activity 3 (See individual student work for correct use of number lines.)

1. 60
2. 120

Lesson 37

PART A

1.	56	56	560	5,600
2.	27	27	270	2,700
3.	24	24	240	2,400

PART B (Items in bold type are given in the text.)

Starting Number	? × 10
50	**5 × 10**
60	6 × 10
70	7 × 10
80	8 × 10
90	9 × 10

Lesson 38

PART A

1.	72	72	720	7,200
2.	21	21	210	2,100
3.	30	30	300	3,000

PART B—Activity 1 (Items in bold type are given in the text.)

Starting Number	? × Largest Power of 10
5,000	**5 × 1,000**
300	3 × 100
6,000	6 × 1,000
700	7 × 100
40	4 × 10
8,000	8 × 1,000

PART B—Activity 2 (Items in bold type are given in the text.)

Number	? × 10	? × 100	? × 1,000
70	**7 × 10**		
80	8 × 10		
900	**90 × 10**	**9 × 100**	
400	40 × 10	4 × 100	
500	50 × 10	5 × 100	
6,000	**600 × 10**	**60 × 100**	**6 × 1,000**
1,000	100 × 10	10 × 100	1 × 1,000
2,000	200 × 10	20 × 100	2 × 1,000
3,000	300 × 10	30 × 100	3 × 1,000

Lesson 39

PART A—Activity 1

1. 42	42	420	4,200
2. 20	20	200	2,000
3. 21	21	210	2,100

PART A—Activity 2 (Items in bold type are given in the text.)

Number	? × 10
70	**7 × 10**
80	8 × 10
90	9 × 10
50	5 × 10
30	3 × 10
40	4 × 10

PART B

1.

	80	6
×		3
		18
+		240
		258

2.

	40	7
×		6
		42
+		240
		282

3.

	30	4
×		5
		20
+		150
		170

4.

	70	8
×		5
		40
+		350
		390

Lesson 40

PART A—Activity 1

1. 6
2. 200
3. 45
4. 450
5. 3
6. 90
7. 900
8. 7
9. 54
10. 540
11. 7
12. 70
13. 64
14. 640
15. 8
16. 160

PART A—Activity 2 (Items in bold type are given in the text.)

Number	? × 10	? × 100	? × 1,000
4,000	**400 × 10**	**40 × 100**	**4 × 1,000**
5,000	500 × 10	50 × 100	5 × 1,000
6,000	600 × 10	60 × 100	6 × 1,000

PART B

1.

	30	7
×		5
		35
+		150
		185

2.

	80	2
×		7
		14
+		560
		574

3.

	60	5
×		3
		15
+		180
		195

PART C

Line A: 8 cm
Line B: 80 mm
Line C: 1 dm
Line D: 15 cm
Line E: 15 mm
Line F: 117 mm
Line G: 10 cm

Lesson 41

PART A

1. 20
2. 24
3. 27
4. 56
5. 200
6. 240
7. 270
8. 56
9. 20
10. 2,400
11. 2,700
12. 560
13. 200
14. 24
15. 27
16. 5,600

PART B (Items in bold type are given in the text.)

Number	? × 10	? × 100	? × 1,000
400	**40 × 10**	**4 × 100**	
500	50 × 10	5 × 100	
8,000	800 × 10	80 × 100	8 × 1,000
60	6 × 10		
6,000	600 × 10	60 × 100	6 × 1,000
90	9 × 10		
500	50 × 10	5 × 100	

PART C

1.
$$\begin{array}{r|r} 40 & 8 \\ \times & 6 \\ \hline & 48 \\ + & 240 \\ \hline & 288 \end{array}$$

3.
$$\begin{array}{r|r} 90 & 3 \\ \times & 4 \\ \hline & 12 \\ + & 360 \\ \hline & 372 \end{array}$$

2.
$$\begin{array}{r|r} 90 & 5 \\ \times & 4 \\ \hline & 20 \\ + & 360 \\ \hline & 380 \end{array}$$

4.
$$\begin{array}{r|r} 80 & 8 \\ \times & 3 \\ \hline & 24 \\ + & 240 \\ \hline & 264 \end{array}$$

Lesson 42

PART A

1. 5
2. 6
3. 9
4. 8
5. 200
6. 4
7. 270
8. 56
9. 7
10. 40
11. 2,700
12. 70

PART B—Activity 1 (See individual student work for correct use of traditional algorithm.)

1. 376
2. 140

PART B—Activity 2

1.
$$\begin{array}{r|r} 40 & 7 \\ \times & 8 \\ \hline & 56 \\ + & 320 \\ \hline & 376 \end{array}$$

2.
$$\begin{array}{r|r} 30 & 5 \\ \times & 4 \\ \hline & 20 \\ + & 120 \\ \hline & 140 \end{array}$$

PART B—Activity 3 (See individual student work for a comparison of the two methods.)

Look for the following types of observations:

- They both involve multiplication and addition.
- They both result in the same answer.
- They both involve regrouping.
- One method is quicker than the other.
- One method uses shorthand.

Lesson 43

PART A

1. 10
2. 100
3. 100
4. 10
5. 10
6. 100
7. 100
8. 10
9. 100
10. 10
11. 10
12. 100

PART B (See individual student work for the correct use of number lines.)

1. Estimation: 29 is about 30 ⟶
   $30 \times 4 = 120$
   Actual Answer: 116
2. Estimation: 219 is about 200 ⟶
   $200 \times 5 = 1{,}000$
   Actual Answer: 1,095
3. Estimation: 792 is about 800 ⟶
   $800 \times 3 = 2{,}400$
   Actual Answer: 2,376
4. Estimation: 209 is about 200 ⟶
   $200 \times 6 = 1{,}200$
   Actual Answer: 1,254

PART C

See individual student work for sketch of the logo. Check lengths to see if they have been measured properly.

Lesson 44

PART A

1. 12
2. 120
3. 1,200
4. 56
5. 56
6. 560
7. 5,600
8. 560
9. 12
10. 120
11. 1,200
12. 30
13. 300
14. 3,000
15. 30
16. 300

PART B

Problems 1 and 3 have errors.

Problem 1: The correct answer is 273. The person who solved the problem carried the 3 instead of the 6.

Problem 3: The correct answer is 534. The person who solved the problem forgot to add the carry.

PART C

Problems 2 and 3 have errors.
Problem 2: The correct answer is 15,371. The student multiplied by 91 instead of by 19.

Problem 3: The correct answer is 480. The student entered too many zeros.

PART D

1. How tall is the goal? (b) 2 meters
   See individual student work for explanation of why the answer must be 2 meters. Look for the following types of observations:
   - 20 cm is too small.
   - 15 mm is too small.
   - 2 m is about 2 yardsticks, and that's about 6 ft.
   - Meters are the biggest measurement there, and the goal is pretty tall.
2. How tall is the elephant? (a) 4 meters
   See individual student work for explanation. Look for the following types of observations:
   - The elephant seems about 2 times taller than the man. An average ht. for a man is 6 ft. That's about 2 m. 2 × 2 = 4 m.
   - 20 meters would be too big.
   - 150 cm would be too small.
3. How tall is the ladder? (a) 10 meters
   See individual student work for explanation. Look for the following types of observations:
   - The ladder is about 5 times bigger than the man. An average ht. for a man is 6 ft. or 2 m. 5 × 2 = 10 m.
   - 40 meters would too big—that's 20 times bigger than the man.
   - 65 meters would also be too big. It's even bigger than 40 m.

Lesson 45

PART A (Items in bold type are given in the text.)

Number	? × 10	? × 100	? × 1,000
200	**20 × 10**	**2 × 100**	**X**
300	30 × 10	3 × 100	X
5,000	500 × 10	50 × 100	5 × 1,000
1,000	100 × 10	10 × 100	1 × 1,000
700	70 × 10	7 × 100	X
400	40 × 10	4 × 100	X
100	10 × 10	1 × 100	X
4,000	400 × 10	40 × 100	4 × 1,000
800	80 × 10	8 × 100	X
10,000	1,000 × 10	100 × 100	10 × 1,000

PART B

See individual student work for Web page design. Be sure students followed the directions in the text.

Lesson 46

PART A

1. 10
2. 100
3. 40
4. 10
5. 10
6. 10
7. 100
8. 10
9. 2
10. 8
11. 10
12. 100

PART B

1. Estimation: 80 × 40 = 3,200
   Calculator Answer: 3,397
2. Estimation: 50 × 20 = 1,000
   Calculator Answer: 1,152
3. Estimation: 20 × 90 = 1,800
   Calculator Answer: 1,869

Lesson 47

PART A

1. 54
2. 35
3. 54
4. 35
5. 540
6. 350
7. 540
8. 350
9. 540
10. 350
11. 540
12. 350

PART B

1. Estimation: 60 × 60 = 3,600
   Calculator Answer: 3,844
2. Estimation: 40 × 40 = 1,600
   Calculator Answer: 1,596
3. Estimation: 70 × 70 = 4,900
   Calculator Answer: 4,964

PART C

See individual student work for the design of the logo. Be sure the triangles match those in the drawing with regard to size and order. The smallest triangle should be on the left, the largest triangle should be next, and so forth.

Lesson 48
PART A
1. 100
2. 1,000
3. 900
4. 240
5. 1,000
6. 700
7. 900
8. 350
9. 100
10. 1,000
11. 540
12. 350

PART B
1. Estimation: $200 \times 30 = 6{,}000$
   Calculator Answer: 6,496
2. Estimation: $600 \times 40 = 24{,}000$
   Calculator Answer: 24,983
3. Estimation: $300 \times 20 = 6{,}000$
   Calculator Answer: 5,661
4. Estimation: $700 \times 30 = 21{,}000$
   Calculator Answer: 21,600
5. Estimation: $300 \times 60 = 18{,}000$
   Calculator Answer: 19,836

Lesson 49
PART A
1. 100
2. 10
3. 10
4. 350
5. 1,000
6. 900
7. 90
8. 300
9. 10
10. 100
11. 200
12. 350

PART B—Activity 1
Traditional Method (See student work for correct use of the traditional algorithm.)

1. 469
2. 324

Expanded Method:

3.
$$\begin{array}{lr|r} & 70 & 8 \\ \times & & 5 \\ \hline & & 40 \\ + & & 350 \\ \hline & & 390 \end{array}$$

4.
$$\begin{array}{lr|r} & 90 & 2 \\ \times & & 8 \\ \hline & & 16 \\ + & & 720 \\ \hline & & 736 \end{array}$$

PART B—Activity 2
1. Estimation: $40 \times 30 = 1{,}200$
   Calculator Answer: 1,216
2. Estimation: $600 \times 50 = 30{,}000$
   Calculator Answer: 27,784
3. Estimation: $200 \times 30 = 6{,}000$
   Calculator Answer: 6,758
4. Estimation: $60 \times 30 = 1{,}800$
   Calculator Answer: 1,638
5. Estimation: $700 \times 10 = 7{,}000$
   Calculator Answer: 7,968

PART C
1. About how long is a paper clip? (b) 5 cm
   See individual student work for explanation.
   Look for the following types of observations:
   - It's about 5 fingertips (or $\frac{1}{2}$ of a computer disk).
   - 5 mm is too small.
   - 50 cm is too big.
2. About how long is a stapler? (b) 20 cm
   See individual student work for explanation.
   Look for the following types of observations:
   - It's about 20 fingertips (or 2 computer disks).
   - 10 m is way too big.
   - 500 cm is the same as 5 m, and that is way too big.
3. About how long is a pencil? (b) 15 cm
   See individual student work for explanation.
   Look for the following types of observations:
   - It's about 15 fingertips (or $1\frac{1}{2}$ computer disks).
   - 15 mm is too small.
   - 15 meters is *really* big!

Lesson 50
PART A
1. 7
2. 30
3. 6
4. 4; 32

PART B
1. 63
2. 48
3. 630
4. 480
5. 6,300
6. 4,800

PART C (Items in bold type are given in the text.)

Starting Number	? × 10
50	**5 × 10**
60	6 × 10
70	7 × 10
80	8 × 10
90	9 × 10

PART D

1.
```
     30 |  7
 ×      |  4
 ------------
          28
 +       120
 ------------
         148
```

2.
```
     90 |  8
 ×      |  6
 ------------
          48
 +       540
 ------------
         588
```

PART E

Number	? × 10	? × 100	? × 1,000
7,000	700 × 10	70 × 100	7 × 1,000
8,000	800 × 10	80 × 100	8 × 1,000
9,000	900 × 10	90 × 100	9 × 1,000

PART F (See individual student work for correct use of the traditional algorithm.)

1. 240
2. 432
3. 672
4. 783

PART G

1. Estimated Extended Fact: 30 × 4 = 120
2. Estimated Extended Fact: 60 × 40 = 2,400
3. Estimated Extended Fact: 400 × 80 = 32,000

PART H

1. Estimation: 80 × 3 = 240
   Calculator Answer: 237
   The error: The person who solved the problem forgot to add the carry.
2. Estimation: 50 × 10 = 500
   Calculator Answer: 672
   The error: The person who solved the problem forgot the place value holder on the second line of the traditional algorithm.
3. Estimation: 100 × 7 = 700
   Calculator Answer: 672
   The error: The person who solved the problem carried the 2 instead of the 4.
4. Estimation: 80 × 9 = 720
   Calculator Answer: 675
   The error: The person who solved the problem carried the 5 instead of the 4.

PART I

1. How long is the bat? (c) 1 meter
   See individual student work for explanation. Look for the following types of observations:
   - We've talked about personal references, and a bat is about a meter.
   - 10 mm is about the width of a fingertip, and that's too small.
   - 1 cm is the same as 10 mm, and again, that's too small.
2. About how long are the skis? (a) 2 meters
   See individual student work for explanation. Look for the following types of observations:
   - The skis are about the length of the man, and an average man is 6 ft. or 2 meters.
   - 12 meters is really long. That's about as tall as 6 men.
   - 20 meters is even longer than 12 meters.
3. About how long is the ship? (b) 100 meters
   See individual student work for explanation. Look for the following types of observations:
   - 100 m is about the size of a football field, and that seems about right.
   - 10 m is too short.
   - 100 cm is only 1 m, and that's too small.

UNIT 4

STUDENT WORKBOOK ANSWER KEY

Division

Lesson 51

PART A

1. 9
2. 3
3. 7
4. 6
5. 7
6. 7
7. 5
8. 6

PART B—Activity 1

1. $6 \times 7 = 42$
   $7 \times 6 = 42$
   $42 \div 6 = 7$
   $42 \div 7 = 6$
2. $5 \times 7 = 35$
   $7 \times 5 = 35$
   $35 \div 7 = 5$
   $35 \div 5 = 7$
3. $8 \times 7 = 56$
   $7 \times 8 = 56$
   $56 \div 7 = 8$
   $56 \div 8 = 7$
4. $8 \times 9 = 72$
   $9 \times 8 = 72$
   $72 \div 9 = 8$
   $72 \div 8 = 9$
5. $3 \times 9 = 27$
   $9 \times 3 = 27$
   $27 \div 9 = 3$
   $27 \div 3 = 9$
6. $3 \times 6 = 18$
   $6 \times 3 = 18$
   $18 \div 6 = 3$
   $18 \div 3 = 6$

PART B—Activity 2

1. $40 \div 5 = 8$ ⟶ 8 piles of cards
2. $9 \times 3 = 27$ ⟶ 27 cards total
3. $6 \times 3 = 18$ ⟶ 18 open seats in the first 6 rows
4. $45 \div 4 = 11$ (with one left over)

PART C

How many square units are in this shape?

- Look for answers that are between 240 and 250. Anything close is acceptable.
- Look for explanations like the following:
  - I drew a big square around the shape and found the area of the square. Then I subtracted the squares and part squares that were not in the shape.
  - I drew a square in the middle part of the shape that was made up of whole squares. I found the area of that. Then I found the area of all the odd-shaped extra squares around the square that I drew. Then I added up all the parts to get the total.

Lesson 52

PART A

1. $6 \times 4 = 24$
   $4 \times 6 = 24$
   $24 \div 6 = 4$
   $24 \div 4 = 6$
2. $8 \times 6 = 48$
   $6 \times 8 = 48$
   $48 \div 6 = 8$
   $48 \div 8 = 6$
3. $5 \times 9 = 45$
   $9 \times 5 = 45$
   $45 \div 9 = 5$
   $45 \div 5 = 9$

PART B (Items in bold type are given in the text.)

Basic Fact	10s Fact	100s Fact
54 ÷ 6 = 9	**540 ÷ 6 = 90**	**5,400 ÷ 6 = 900**
81 ÷ 9 = 9	810 ÷ 9 = 90	**8,100 ÷ 9 = 900**
36 ÷ 4 = 9	**360 ÷ 4 = 90**	3,600 ÷ 4 = 900
24 ÷ 4 = 6	240 ÷ 4 = 60	**2,400 ÷ 4 = 600**
72 ÷ 8 = 9	720 ÷ 8 = 90	**7,200 ÷ 8 = 900**
12 ÷ 4 = 3	**120 ÷ 4 = 30**	1,200 ÷ 4 = 300

PART C

1. 3
2. 6
3. 80
4. 30
5. 9
6. 9
7. 80
8. 90

PART D

Shape A: The area of the triangle is exactly 30. Acceptable answers—28 to 32.

Shape B: The area of the rectangle with one jagged side should be between 34 and 37.

Shape C: The area of the "blob" should be between 20 and 28.

Shape D: The area of the tall rectangle is 18. This can be computed exactly.

Look for an explanation like the following: The Shape B is the largest. I counted all the squares, parts, and wholes, of all the shapes and I found B to be the largest—36.

Lesson 53

PART A

1. 2
2. 90
3. 70
4. 8
5. 40
6. 100

PART B (See individual student work for correct use of number lines.)

1. 3
2. 4
3. 4
4. 4
5. 2

Lesson 54

PART A (See individual student work for correct use of number lines.)

1. 2
2. 3
3. 5
4. 4

PART B (Items in bold type are given in the text.)

Basic Fact	10s Fact	100s Fact
72 ÷ 8 = 9	**720 ÷ 8 = 90**	**7,200 ÷ 8 = 900**
49 ÷ 7 = 7	490 ÷ 7 = 70	**4,900 ÷ 7 = 700**
25 ÷ 5 = 5	250 ÷ 5 = 50	**2,500 ÷ 5 = 500**
24 ÷ 6 = 4	**240 ÷ 6 = 40**	2,400 ÷ 6 = 400
42 ÷ 7 = 6	420 ÷ 7 = 60	4,200 ÷ 7 = 600
27 ÷ 9 = 3	270 ÷ 9 = 30	**2,700 ÷ 9 = 300**
16 ÷ 4 = 4	160 ÷ 4 = 40	**1,600 ÷ 4 = 400**

PART C

Area of Object A: Acceptable range of answers is 60 to 70 square units (best answer: 62 or 63).

Area of Object B: Acceptable range of answers is 15 to 20 square units (best answer: 16, 17, or 18).

Why do the measurements of the objects appear to be different?

Look for answers that incorporate reasons like the following:

- Object A is drawn on a grid with smaller squares.
- Object B is drawn on a grid with bigger squares.
- It takes about 4 squares in A to make up 1 square in B.

Lesson 55

PART A

1. 3
2. 50
3. 6
4. 60

PART B—Activity 1

1. 8.5 (a) 8 (b) .5
2. 6.5 (a) 6 (b) .5

PART B—Activity 2 (See individual student work for correct use of number lines.)

1. 7 R2
2. 6 R5
3. 6 R3
4. 4 R4

PART C

What is the approximate population of San Francisco? <u>818,400</u>

How did you get your answer?

Student responses should include these types of discussions:

- I drew a big square around the map and found the area. It was 6 × 7 = 42. Then I subtracted all the whole squares outside the map. Then I subtracted all the pieces of squares outside the map (I put them together to make whole squares out of them). Then I got 31. Then I multiplied 31 × 26,400 people. Then I got the final answer of 818,400.
- I counted all the whole squares in the map. Then I looked at the parts of squares and pieced them together to make whole squares. Then I added those numbers together and got a total of 31. Then I multiplied by 26,400 people. Then I got the final answer of 818,400.

Lesson 56

PART A

1. 3.166666667
2. 3.111111111
3. 6.25
4. 6.166666667

PART B

1. Calculator Answer: 4.25
   Rounded Answer: 4
2. Calculator Answer: 8.285
   Rounded Answer: 8
3. Calculator Answer: 8.2
   Rounded Answer: 8
4. Calculator Answer: 5.33
   Rounded Answer: 5

Lesson 57

PART A

1. Calculator Answer: 2.1666667
   Rounded Answer: 2
2. Calculator Answer: 9.44444
   Rounded Answer: 9
3. Calculator Answer: 6.25
   Rounded Answer: 6
4. Calculator Answer: 6.33333
   Rounded Answer: 6

PART B

1. $\begin{array}{r}6\\2\overline{)12}\end{array}$
2. $\begin{array}{r}9\\9\overline{)81}\end{array}$
3. $\begin{array}{r}6\\7\overline{)42}\end{array}$
4. $\begin{array}{r}4\\8\overline{)32}\end{array}$
5. $\begin{array}{r}8\\4\overline{)32}\end{array}$
6. $\begin{array}{r}7\\5\overline{)35}\end{array}$
7. $\begin{array}{r}10\\2\overline{)20}\end{array}$
8. $\begin{array}{r}8\\9\overline{)72}\end{array}$

PART C

1. 20 sq. units
2. about $(20 \times 5) = 100 + 20 = 120$ sq. units
3. about 6 times or $120 - 20 = 100$ sq. units
4. about $50 \times 5 = 250 + 120 = 370$ sq. units
5. Student responses should include something like this: I found the area of several parts and added them together to get the total area.

Lesson 58

PART A

1. 7
2. 7
3. 72

PART B

1. 70
2. 70
3. 90

PART C

1. $7 \times 4 = 28$
   $4 \times 7 = 28$
   $28 \div 4 = 7$
   $28 \div 7 = 4$
2. $8 \times 7 = 56$
   $7 \times 8 = 56$
   $56 \div 7 = 8$
   $56 \div 8 = 7$
3. $4 \times 3 = 12$
   $3 \times 4 = 12$
   $12 \div 3 = 4$
   $12 \div 4 = 3$

PART D (See individual student work for correct use of number lines.)

1. 4
2. 3

PART E

1. Calculator Answer: 3.11111
   Rounded Answer: 3
2. Calculator Answer: 7.375
   Rounded Answer: 7
3. Calculator Answer: 5.285714286
   Rounded Answer: 5
4. Calculator Answer: 9.5
   Rounded Answer: 10

PART F

1. $\begin{array}{r}7\\6\overline{)42}\end{array}$ or $\begin{array}{r}8\\6\overline{)48}\end{array}$
2. $\begin{array}{r}8\\9\overline{)72}\end{array}$ or $\begin{array}{r}9\\9\overline{)81}\end{array}$

Lesson 59

PART A—Activity 1

1. 3
2. 4
3. 6
4. 6
5. 5
6. 7
7. 7
8. 10

PART A—Activity 2

1. $\begin{array}{r}7\\7\overline{)49}\end{array}$
2. $\begin{array}{r}3\\9\overline{)27}\end{array}$
3. $\begin{array}{r}9\\4\overline{)36}\end{array}$
4. $\begin{array}{r}6\\8\overline{)48}\end{array}$

PART B

1. 70.5
2. 31.6
3. 100.14
4. 20.57
5. 70
6. 90.44

PART C

1. 398 sq. units

2a. kitchen

2b. bathroom

2c. about 2 times bigger or 88 − 42 = 46 sq. units difference

2d. Bedroom B is 63 sq. units and Bedroom A is 77 sq. units. 77 − 63 = 14 sq. units. Bedroom A is 14 sq. units bigger than Bedroom B.

Lesson 60

PART A—Activity 1 (Items in bold type are given in the text.)

Basic Fact	10s Fact	100s Fact
54 ÷ 6 = 9	**540 ÷ 6 = 90**	**5,400 ÷ 6 = 900**
64 ÷ 8 = 8	640 ÷ 8 = 80	**6,400 ÷ 8 = 800**
36 ÷ 9 = 4	**360 ÷ 9 = 40**	3,600 ÷ 9 = 400
18 ÷ 9 = 2	180 ÷ 9 = 20	**1,800 ÷ 9 = 200**

PART A—Activity 2

1. 59
2. 74

PART B

1. $\begin{array}{r}20\\6\overline{)120}\end{array}$

2. $\begin{array}{r}20\\9\overline{)180}\end{array}$

3. $\begin{array}{r}50\\4\overline{)200}\end{array}$

4. $\begin{array}{r}70\\5\overline{)350}\end{array}$

5. $\begin{array}{r}70\\7\overline{)490}\end{array}$

PART C

1. What is the size of each object? (See individual student work. Allow some flexibility if the answers are close.)
   - Bed: 50 sq. units
   - Closet: 10 sq. units
   - Rug: acceptable range 75 – 80 sq. units
   - Desk: 12 sq. units
2. What is the total size of the bedroom in the drawing?
   20 × 20 = 400 sq. units
3. Is there room for a 9 x 12 bed without covering any part of the rug?
   No.

Lesson 61

PART A—Activity 1

1. $\begin{array}{r}70\\7\overline{)490}\end{array}$

2. $\begin{array}{r}70\\6\overline{)420}\end{array}$

PART A—Activity 2

1. 6 × 7 = 42
   7 × 6 = 42
   42 ÷ 7 = 6
   42 ÷ 6 = 7
2. 9 × 3 = 27
   3 × 9 = 27
   27 ÷ 9 = 3
   27 ÷ 3 = 9
3. 6 × 9 = 54
   9 × 6 = 54
   54 ÷ 9 = 6
   54 ÷ 6 = 9

PART B

1. $\begin{array}{r}7\\80\overline{)560}\end{array}$

2. $\begin{array}{r}6\\60\overline{)360}\end{array}$

3. $\begin{array}{r}6\\50\overline{)300}\end{array}$

4. $\begin{array}{r}10\\40\overline{)400}\end{array}$

PART C

1. $9 \times 10\overline{)36 \times 10}$

2. $7 \times 10\overline{)49 \times 10}$

3. $6 \times 10\overline{)24 \times 10}$

4. $8 \times 10\overline{)56 \times 10}$

5. $3 \times 10\overline{)9 \times 10}$

PART D

See individual student work. Look for at least four items. Check the table to be sure the items are all the right sizes and will fit in the room.

Lesson 62

PART A—Activity 1

1. 9
2. 7
3. 8
4. 4
5. 6
6. 9

PART A—Activity 2

1. $9 \times 10\overline{)81 \times 10}$

2. $7 \times 10\overline{)21 \times 10}$

3. $8 \times 10\overline{)72 \times 10}$

4. $6 \times 10\overline{)48 \times 10}$

PART B (Items in bold type are given in the text.)

Near Extended Fact	Pull Out the 10s	Basic Fact	Near Extended Fact and Solution
$60\overline{)240}$	$6 \times 10\overline{)24 \times 10}$	$\begin{array}{r}4\\6\overline{)24}\end{array}$	$\begin{array}{r}4\\60\overline{)240}\end{array}$
$70\overline{)280}$	$7 \times 10\overline{)28 \times 10}$	$\begin{array}{r}4\\7\overline{)28}\end{array}$	$\begin{array}{r}4\\70\overline{)280}\end{array}$
$30\overline{)210}$	$3 \times 10\overline{)21 \times 10}$	$\begin{array}{r}7\\3\overline{)21}\end{array}$	$\begin{array}{r}7\\30\overline{)210}\end{array}$
$50\overline{)250}$	$5 \times 10\overline{)25 \times 10}$	$\begin{array}{r}5\\5\overline{)25}\end{array}$	$\begin{array}{r}5\\50\overline{)250}\end{array}$
$20\overline{)180}$	$2 \times 10\overline{)18 \times 10}$	$\begin{array}{r}9\\2\overline{)18}\end{array}$	$\begin{array}{r}9\\20\overline{)180}\end{array}$
$80\overline{)320}$	$8 \times 10\overline{)32 \times 10}$	$\begin{array}{r}4\\8\overline{)32}\end{array}$	$\begin{array}{r}4\\80\overline{)320}\end{array}$

Lesson 63

PART A—Activity 1

1. 8
2. 6
3. 7
4. 3
5. 9
6. 10

PART A—Activity 2

1. $2 \times 10\overline{)12 \times 10}$ $\begin{array}{r}6\\2\overline{)12}\end{array}$
2. $8 \times 10\overline{)64 \times 10}$ $\begin{array}{r}8\\8\overline{)64}\end{array}$
3. $4 \times 10\overline{)28 \times 10}$ $\begin{array}{r}7\\4\overline{)28}\end{array}$
4. $7 \times 10\overline{)63 \times 10}$ $\begin{array}{r}9\\7\overline{)63}\end{array}$

PART B

1. Near Fact: $28 \div 7 = 4$
2. Exact Answer: 4.428571
3. The student switched the numbers around and divided 7 by 31. We know this because the answer starts with a decimal. That situation happens only when we divide a smaller number by a bigger number.

PART C

Problems B and C have errors.

In Problem B, the student lined up the place values incorrectly. The correct answer should be 69 R4. A good way to check for this is to do an estimate. Think of a near fact, like $350 \div 5 = 70$. The answer of 6 R4 is way too low. This is usually a place value problem.

In Problem C, the student lined up the place values incorrectly. The correct answer should be 40 R1. A good way to check for this is to do an estimate. Think of a near fact, like $120 \div 3 = 40$. The answer of 4 R1 is obviously too low. This usually means that there was a place value error made in the calculations.

Lesson 64

PART A—Activity 1

1. 5
2. 10
3. 10
4. 7
5. 10
6. 6

PART A—Activity 2

1. $30\overline{)270}$
2. $60\overline{)720}$
3. $70\overline{)490}$

PART B

1. Near Fact: $200 \div 20 = 10$
2. Exact Answer: 9.142857
3. The student switched the digits around in the number 192. He or she divided 129 by 21. That would result in an answer of 6 something.

PART C

1. Large triangles: 18 squares each
   $18 \times 4 = 72$ triangles each
2. Small triangles: 8 squares each
   $8 \times 4 = 32$ triangles each
3. Small squares: 4 squares each
   $4 \times 4 = 16$ triangles each

Lesson 65

PART A

1. 7	70	700
2. 7	70	700
3. 8	80	800
4. 3	30	300

PART B

1. $9 \times 7 = 63$
   $7 \times 9 = 63$
   $63 \div 9 = 7$
   $63 \div 7 = 9$
2. $6 \times 8 = 48$
   $8 \times 6 = 48$
   $48 \div 6 = 8$
   $48 \div 8 = 6$
3. $5 \times 7 = 35$
   $7 \times 5 = 35$
   $35 \div 7 = 5$
   $35 \div 5 = 7$

PART C (See individual student work for correct use of number lines.)

1. 3
2. 4
3. 20

PART D

1. The near fact is $42 \div 6 = 7$.
2. The near fact is $49 \div 7 = 7$.

PART E

1. 62
2. 72

PART F

The near extended fact is $240 \div 30 = 8$.
The near extended fact is $280 \div 40 = 7$.

PART G (Items in bold type are given in the text.)

Near Extended Fact	Pull Out the 10s	Basic Fact
$60\overline{)240}$	**$6 \times 10\overline{)24 \times 10}$**	**$\overset{4}{6\overline{)24}}$**
$80\overline{)720}$	$8 \times 10\overline{)72 \times 10}$	$\overset{9}{8\overline{)72}}$
$50\overline{)250}$	$5 \times 10\overline{)25 \times 10}$	$\overset{5}{5\overline{)25}}$
$40\overline{)160}$	$4 \times 10\overline{)16 \times 10}$	$\overset{4}{4\overline{)16}}$

PART H

1. A near fact is $56 \div 8 = 7$
2. The exact answer is 7.375.
3. The student switched the numbers around and divided 8 by 59. The reason I think that is that the answer given starts with a decimal. This happens only when you divide a smaller number by a larger number.

STUDENT WORKBOOK ANSWER KEY

Factors, Patterns, and Multiples

UNIT 5

Lesson 66

PART A

1. 13
2. 12
3. 16
4. 48
5. 6
6. 48
7. 8
8. 8
9. 4

PART B

1 2 3 4 5 6 7 8 9 10 11 12 13 14 15

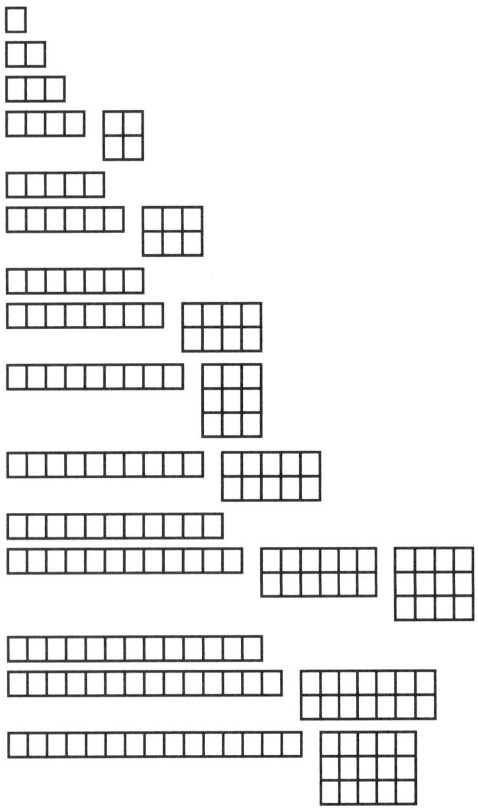

PART B

16 17 18 19 20 21 22 23 24 25

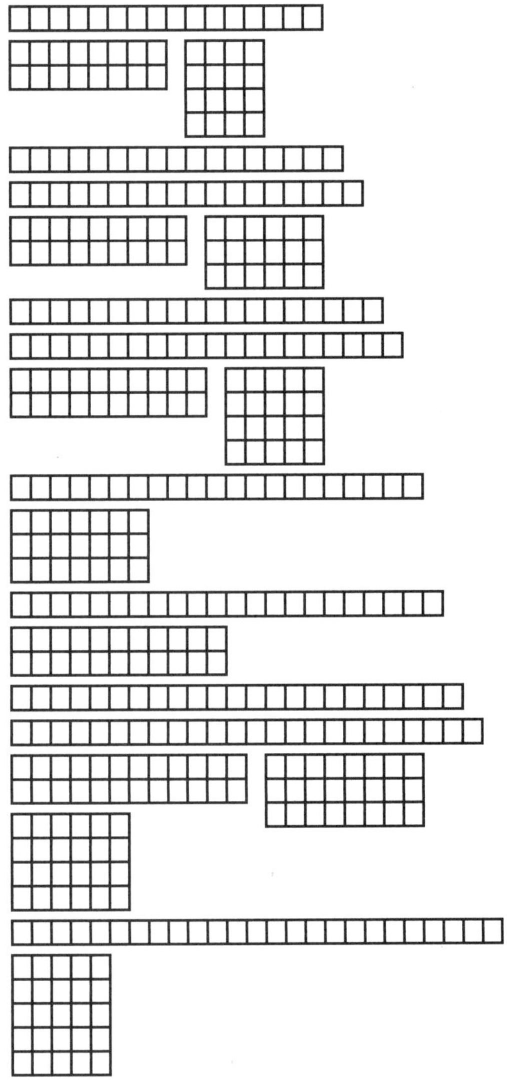

Lesson 67

PART A

1. 9
2. 120
3. 1,100
4. 30
5. 90
6. 9
7. 800
8. 9
9. 6

Lesson 68

PART A

1. 12
2. 12
3. 18
4. 18
5. 24
6. 24

PART B

(Items in bold type are given in the text.)

Number	Multiplication Problems	Factors
1	**1 × 1 = 1**	**1**
2	**1 × 2 = 2**	**1, 2**
3	**1 × 3 = 3**	**1, 3**
4	**1 × 4 = 4** **2 × 2 = 4**	**1, 2, 4**
5	1 × 5 = 5	1, 5
6	1 × 6 = 6 2 × 3 = 6	1, 2, 3, 6
7	1 × 7 = 7	1, 7
8	1 × 8 = 8 2 × 4 = 8	1, 2, 4, 8
9	1 × 9 = 9 3 × 3 = 9	1, 3, 9
10	1 × 10 = 10 2 × 5 = 10	1, 2, 5, 10
11	1 × 11 = 11	1, 11
12	**1 × 12 = 12** **2 × 6 = 12** **3 × 4 = 12**	**1, 2, 3,** **4, 6, 12**
13	1 × 13 = 13	1, 13
14	1 × 14 = 14 2 × 7 = 14	1, 2, 7, 14
15	**1 × 15 = 15** **3 × 5 = 15**	**1, 3, 5, 15**
16	1 × 16 = 16 2 × 8 = 16 4 × 4 = 16	1, 2, 4, 8, 16
17	1 × 17 = 17	1, 17
18	1 × 18 = 18 2 × 9 = 18 3 × 6 = 18	1, 2, 3, 6, 9, 18
19	1 × 19 = 19	1, 19
20	1 × 20 = 20 2 × 10 = 20 4 × 5 = 20	1, 2, 4, 5, 10, 20
21	1 × 21 = 21 3 × 7 = 21	1, 3, 7, 21
22	1 × 22 = 22 2 × 11 = 22	1, 2, 11, 22
23	1 × 23 = 23	1, 23
24	1 × 24 = 24 2 × 12 = 24 3 × 8 = 24 4 × 6 = 24	1, 2, 3, 4, 6, 8, 12, 24
25	1 × 25 = 25 5 × 5 = 25	1, 5, 25

PART C

(Answers will vary depending on the rectangles drawn by students but should be close to the answers given here.)

Area of A = 77 (acceptable range 77–84)
Area of B = 88 (acceptable range 77–88)
Area of C = 84 (acceptable range 66–84)
Area of D = 24 (acceptable range 24–35)

Lesson 69

PART A

1. 24
12
3
6

2. 1 × 16 = 16
2 × 8 = 16
4 × 4 = 16

3. 1 × 15 = 15
3 × 5 = 15

4. 1 × 18 = 18
2 × 9 = 18
3 × 6 = 18

PART B

1. Factor rainbow for 16.

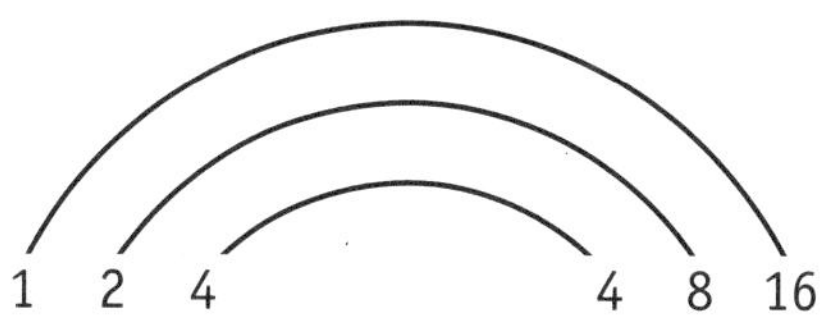

The factors of 16 are 1, 2, 4, 8, and 16.

2. Factor rainbow for 15.

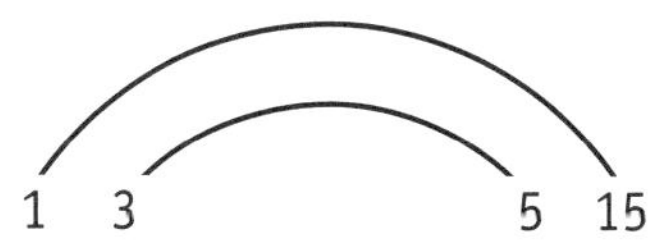

The factors of 15 are 1, 3, 5, and 15.

3. Factor rainbow for 12.

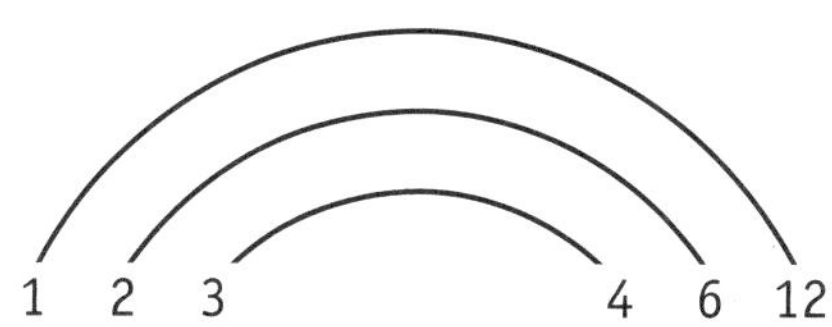

The factors of 12 are 1, 2, 3, 4, 6, and 12.

4. Factor rainbow for 7.

The factors of 7 are 1 and 7.

PART C

Look at individual student work. Be sure to check for the following:

1. Four rectangles labeled B, C, D, and E.
2. Correct multiplication.
3. Correct units (square units).

Lesson 70

PART A

1. 1, 3, and 9 should be circled.
2. 1, 2, 3, 4, 6, and 12 should be circled.
3. 1, 3, 5, and 15 should be circled.
4. 1 and 7 should be circled.
5. 1, 3, 7, and 21 should be circled.

PART B

1. Factor rainbow for 25.

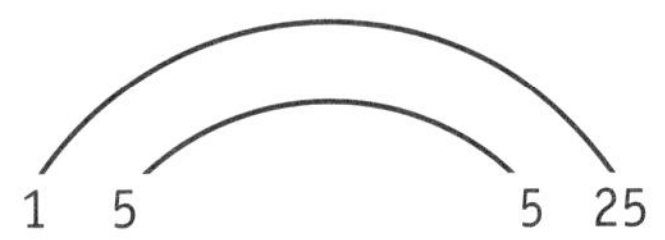

The factors of 25 are 1, 5, and 25 (remember, no duplicates).

2. Factor rainbow for 27.

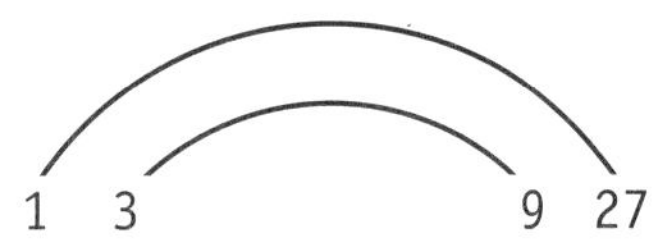

The factors of 27 are 1, 3, 9, and 27.

3. Factor rainbow for 32.

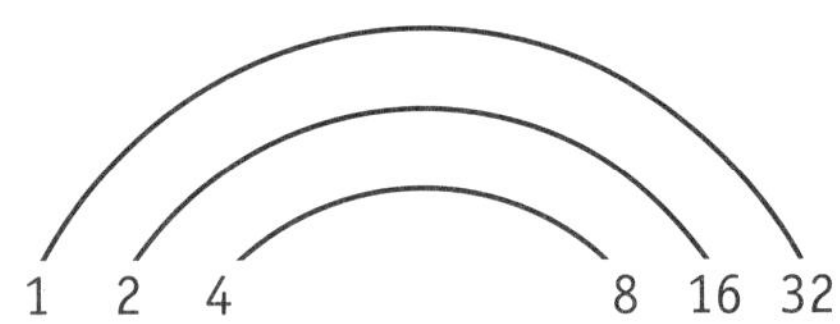

The factors of 32 are 1, 2, 4, 8, 16, and 32.

4. Factor rainbow for 36.

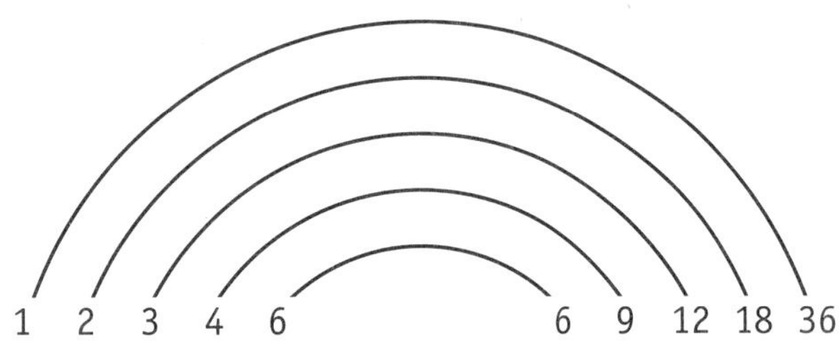

The factors of 36 are 1, 2, 3, 4, 6, 9, 12, 18, and 36.

Lesson 71

PART A

1. The factors of 27 are 1, 3, 9, and 27.
2. The factors of 32 are 1, 2, 4, 8, 16, and 32.
3. The factors of 40 are 1, 2, 4, 5, 8, 10, 20, and 40.
4. The factors of 9 are 1, 3, and 9.

PART B—Activity 1

Students should have circled the following numbers: 2, 3, 5, 7, 11, 13, 17, 19, and 23.

PART B—Activity 2

1. Factor rainbow for 17.

The factors of 17 are 1 and 17.

2. Factor rainbow for 23.

The factors of 23 are 1 and 23.

3. Factor rainbow for 31.

The factors of 31 are 1 and 31.

4. Look at individual student work. Be sure that the students' descriptions include one or more of the following observations:
   - A prime number has only two factors.
   - The factors are 1 and the number itself.
   - There is only one "arc" in the rainbow.

PART C

1. Tomatoes, 26 meters; potatoes, 42 meters; corn, 44 meters; peas, 18 meters; carrots, 26 meters
2. 18 meters
3. 52 meters

Lesson 72

PART A

1. 90	**4.** 400	**7.** 80
2. 70	**5.** 60	**8.** 80
3. 700	**6.** 800	**9.** 40

PART B—Activity 1

See individual student work. Look for these kinds of observations:

1. Most of them are odd numbers.
2. Only one of them is an even number, and that's 2.
3. Most of them end with 1, 3, 7, or 9.
4. None of them end with 4, 6, 8, or 0.
5. Only one of them ends in a 2, and that's 2.

PART B—Activity 2

See individual student work. Look for these kinds of observations:

1. I crossed out the evens because they all have 2 as a factor.
2. I crossed out all the numbers that end in 5 because they have 5 as a factor.
3. I crossed out all the numbers that end in 0 because they have 10 as a factor.
4. I crossed out the 4s because they have 2 as a factor.
5. I didn't cross out numbers ending in 1, 3, 7, or 9 because in the last activity, most of the primes ended in those numbers.

Look at students' guesses about prime numbers and see how many were correct.

Lesson 73

PART A—Activity 1

1. 5	**4.** 30	**7.** 500
2. 1,200	**5.** 150	**8.** 100
3. 140	**6.** 90	**9.** 6

PART A—Activity 2

The following numbers should be circled: 2, 3, 5, 11, 13, 29, and 31.

PART A—Activity 3

The factors of 12 are 1, 2, 3, 4, 6, and 12.
The factors of 9 are 1, 3, and 9.
The factors of 13 are 1 and 13.
The factors of 25 are 1, 5, and 25.

PART B—Activity 1

See individual student responses. Sample response: It looks like the triangle is half of the rectangle so the area is probably half of the area of the rectangle.

PART B—Activity 2

See individual student responses. Sample response: The parallelogram has a rectangle inside of it. I can find the area of that first. Then, when I look at the extra parts, I see that they form two triangles that can be put together to make another rectangle. I can find the area of that and add the two areas together. Another acceptable response is: I took the long side and multiplied it by the short side because you can make a big rectangle if you rearrange the parts of the parallelogram.

Lesson 74

PART A

The factors of 17 are 1 and 17.
The factors of 2 are 1 and 2.
The factors of 24 are 1, 2, 3, 4, 6, 8, 12, and 24.
The factors of 8 are 1, 2, 4, and 8.
The factors of 6 are 1, 2, 3, and 6.

PART B—Activity 1

The primes are 2, 5, and 11.

PART B—Activity 2

The numbers not circled are 1, 4, 8, 15, 21, and 49.

PART B—Activity 3

1 is not prime or composite. It's a special case, and it does not have a factor tree.

The prime factor tree for 4 is:

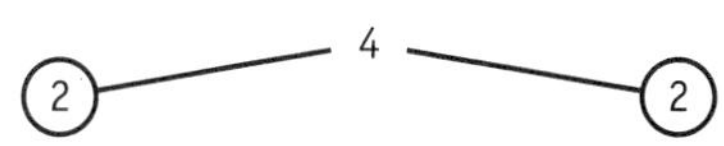

The prime factor tree for 8 is:

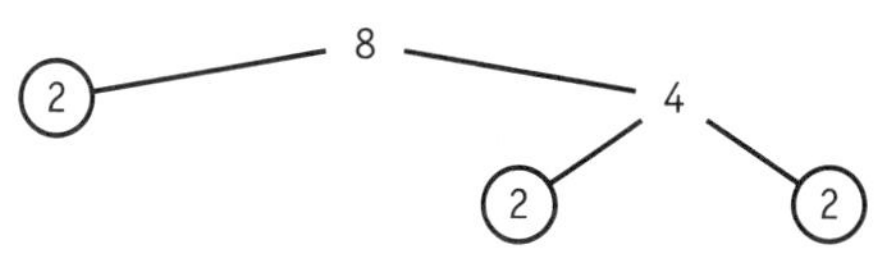

The prime factor tree for 15 is:

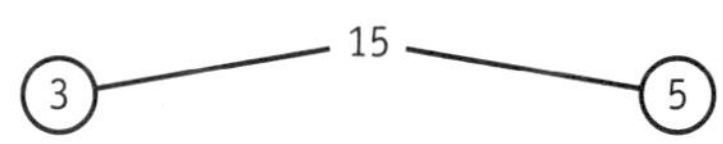

The prime factor tree for 21 is:

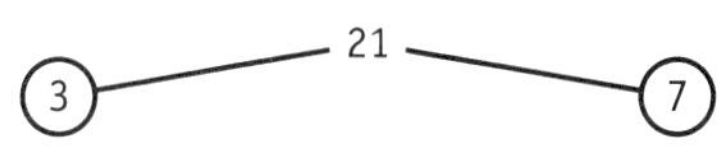

The prime factor tree for 49 is:

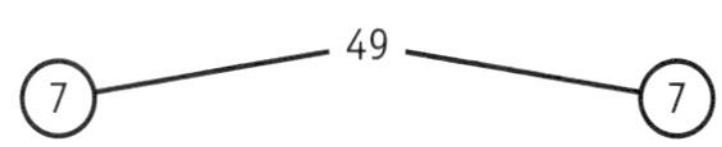

PART C

1. 363 square units
2. 36 square units

Lesson 75

PART A—Activity 1

1. 9	**3.** 28	**5.** 4
2. 42	**4.** 4	**6.** 56

PART A—Activity 2

The factors of 45 are 1, 3, 5, 9, 15, 45.
The factors of 28 are 1, 2, 4, 7, 14, 28.
The factors of 42 are 1, 2, 3, 6, 7, 14, 21, 42.
The factors of 20 are 1, 2, 4, 5, 10, 20.
The factors of 36 are 1, 2, 3, 4, 6, 9, 12, 18, 36.

PART B

The prime factor tree for 20 is:

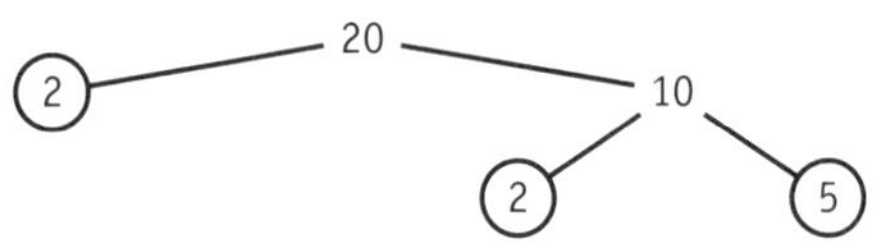

The prime factor tree for 28 is:

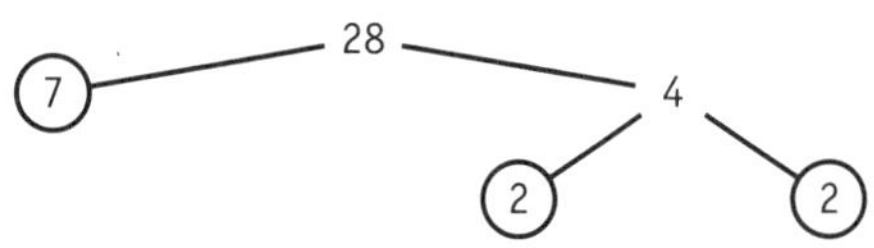

PART C—Activity 1

See individual student work. Look for the following:

1. Three rectangles of different sizes.
2. The area and perimeter for each.
3. Correct units (square units for area, units for perimeter).
4. Correct multiplication and/or addition.

PART C—Activity 2

See individual student work. Look for the following:

1. Depending on the rectangles drawn, some students will respond that the area is not always bigger than the perimeter.
2. Depending on the rectangles drawn, some students will respond that the area is always bigger than the perimeter. In this case, you need to provide examples showing that that is not always true.

Lesson 76

PART A

The prime factor tree for 21 is:

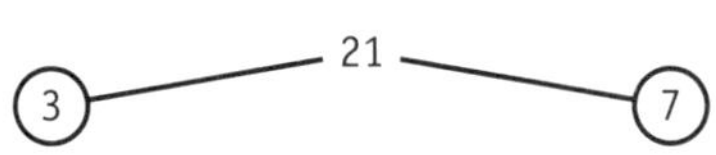

The prime factorization of 21 is 3×7.

The prime factor tree for 40 is:

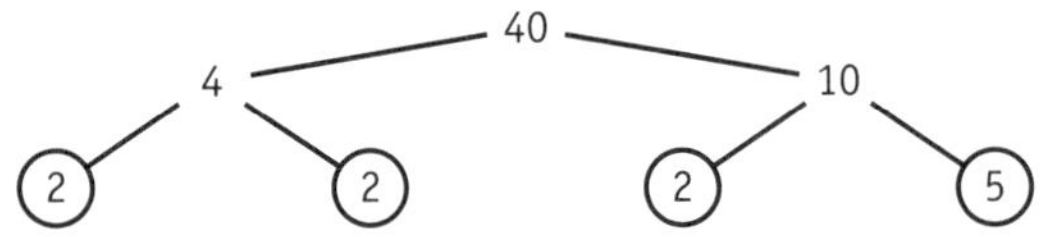

The prime factorization of 40 is $2 \times 2 \times 2 \times 5$.
The prime factor tree for 35 is:

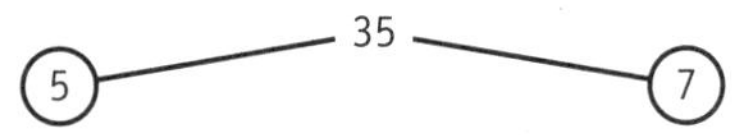

The prime factorization of 35 is 5×7.
The prime factor tree for 60 is:

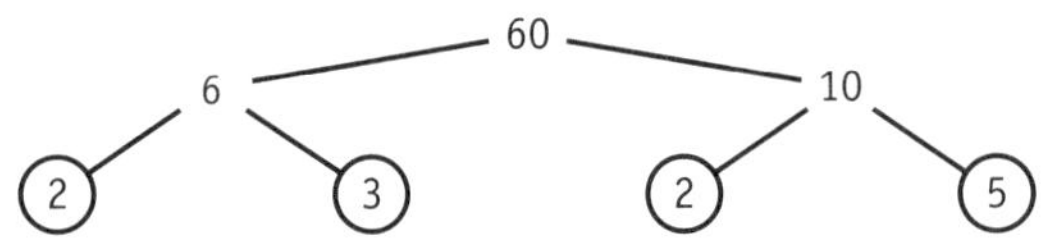

The prime factorization of 60 is $2 \times 2 \times 3 \times 5$.
The prime factor tree for 70 is:

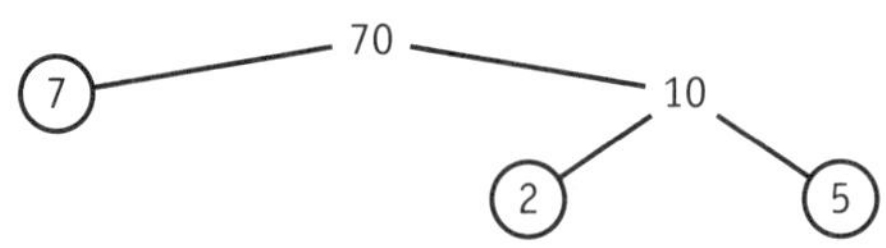

The prime factorization of 70 is $2 \times 5 \times 7$.
The prime factor tree for 80 is:

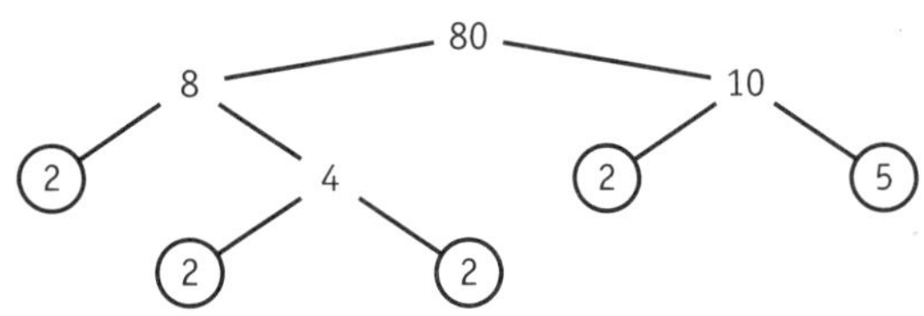

The prime factorization of 80 is $2 \times 2 \times 2 \times 2 \times 5$.

PART B

See individual student work. Look for the following:

1. Several rectangles with perimeter of 40 but with different areas.
2. The area written inside each of the rectangles in the format A = ____. (The perimeter does not need to be written down because it should be 40 for every rectangle.)
3. The area labeled with square units.

Lesson 77

PART A

1. 42, 42, 420, 4,200
2. 20, 20, 200, 2,000
3. 21, 21, 210, 2,100

PART B—Activity 1

(Items in bold type are given in the text.)

Rectangle	Base	Height	Difference	Perimeter	Area
1	**19**	**1**	**18**	**40**	**19**
2	18	2	16	40	36
3	17	3	14	40	51
4	16	4	12	40	64
5	15	5	10	40	75
6	14	6	8	40	84
7	13	7	6	40	91
8	12	8	4	40	96
9	11	9	2	40	99
10	**10**	**10**	**0**	**40**	**100**

PART B—Activity 2

Graph A

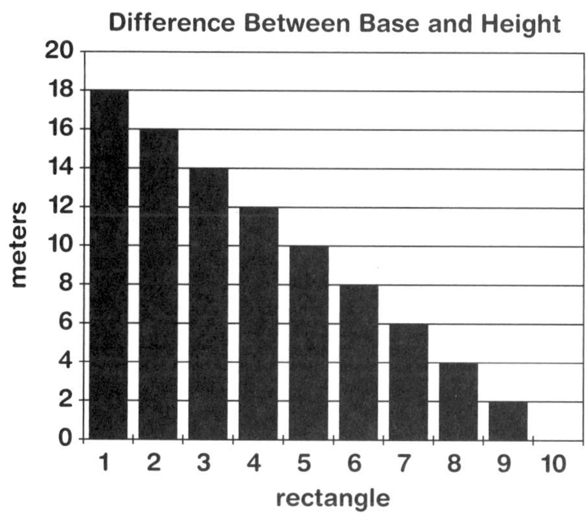

Graph B

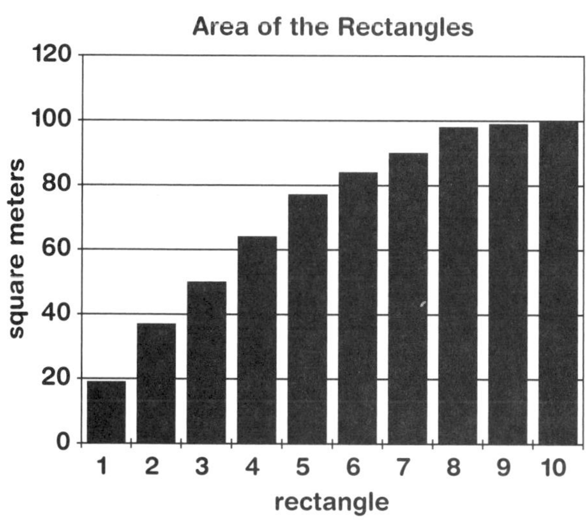

See individual work in the bar charts. Be sure the students have graphed the right information. The trend for Graph A should be decreasing numbers. The trend for Graph B should be increasing numbers. The conclusion is that the area increases as the difference between height and base becomes smaller.

Answer to Questions:

1. The area increases as the difference between the height and base gets smaller.
2. The one that has the same height and base (or a square, in other words).

Lesson 78

PART A

The arrays for 12 are:

12 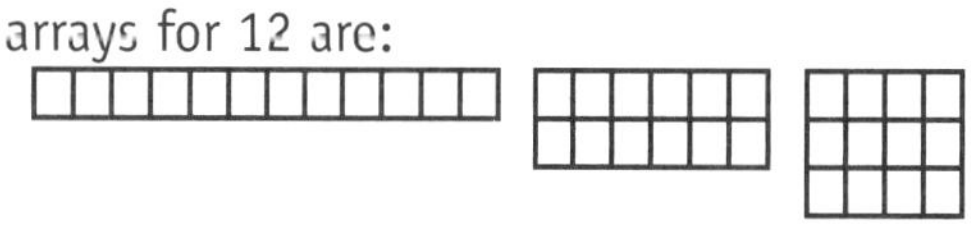

The arrays for 17 are:

17 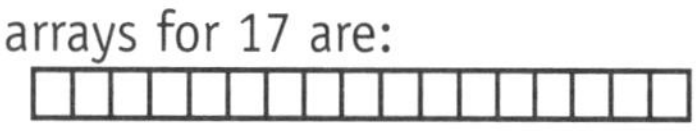

The arrays for 25 are:

25 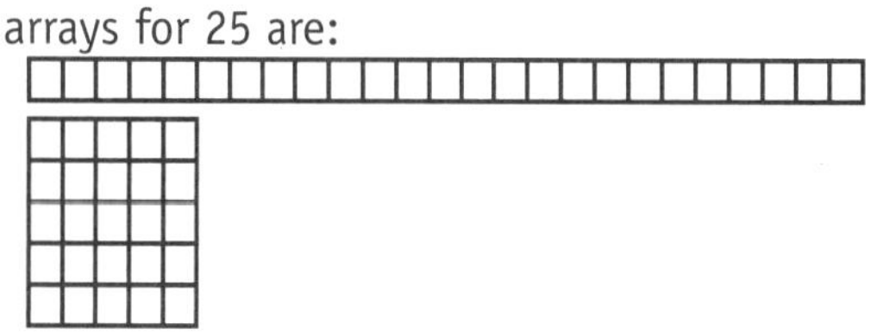

PART B

1. The factors of 16 are 1, 2, 4, 8, and 16.
2. The factors of 9 are 1, 3, and 9.
3. The factors of 21 are 1, 3, 7, and 21.
4. The factors of 31 are 1 and 31.

PART C

1. The factor rainbow for 12 is:

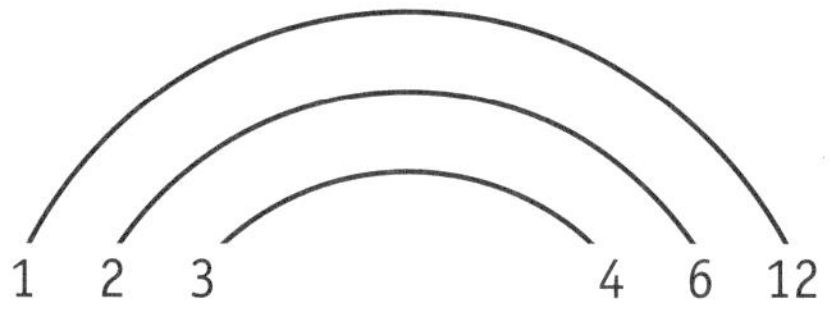

The factors of 12 are 1, 2, 3, 4, 6, and 12.

2. The factor rainbow for 15 is:

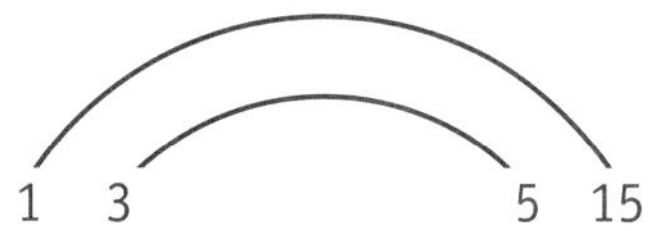

The factors of 15 are 1, 3, 5, and 15.

3. The factor rainbow for 17 is:

The factors of 17 are 1 and 17.

4. The factor rainbow for 25 is:

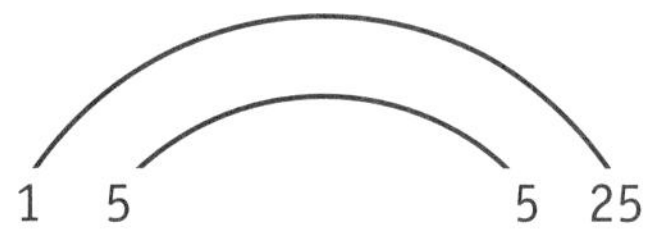

The factors of 25 are 1, 5, and 25 (remember, no duplicates).

PART D

1. The area is $2 \times 4 = 8$ sq. cm. The perimeter is $2 + 2 + 4 + 4 = 12$ cm.
2. The area is $3 \times 3 = 9$ sq. in. The perimeter is $3 + 3 + 3 + 3 = 12$ in.
3. The area is $4 \times 10 = 40$ sq. m. The perimeter is $4 + 4 + 10 + 10 = 28$ m.

PART E

1. Composite
2. Composite
3. Prime
4. Composite
5. Prime
6. Prime
7. Composite

PART F

Area of A = $3 \times 6 = 18$ sq. units.
Area of B = $2 \times 8 = 16$ sq. units.
Area of C = half of 8×3 = half of 24 = 12 sq. units.
Area of D = $7 \times 4 = 28$ sq. units.

PART G

1. Circles from left to right: 2, 3, 2, 3
   Prime factorization of 36 = $2 \times 2 \times 3 \times 3$
2. Circles from left to right: 5, 5
   Prime factorization of 50 = $2 \times 5 \times 5$
3. Circles from left to right: 2, 3, 3, 3
   Prime factorization of 54 = $2 \times 3 \times 3 \times 3$

Lesson 79

PART A

1. 2
2. 15
3. 9
4. 7
5. 5
6. 12

PART B—Activity 1

(Items in bold type are given in the text.)

Number	Divide by 2	Divide by 5	Divide by 10
10	**X**	**X**	**X**
12	X		
15		X	
27			
65		X	
150	X	X	X

1. Example of a five-digit number that you can divide by 2: 12,334
2. Example of a five-digit number that you can divide by 5: 12,335
3. Example of a five-digit number that you can divide by 10: 12,330

PART B—Activity 2

(Items in bold type are given in the text.)

Number	Divide by 3	Divide by 6
6	**X**	**X**
12	X	X
15	X	
21	X	
24	X	X
51	X	

1. Example of four-digit number that you can divide by 3: 4,122
2. Example of a four-digit number that you can divide by 6: 1,242

PART C

1. 4 square units, 16 square units
2. 16 square units
3. 36 square units
4. 68 square units

Lesson 80
PART A

1. 4
2. 1,000
3. 6
4. 6
5. 2
6. 500
7. 50
8. 4
9. 200

PART B
The rules that apply to 555 are the 3 and 5 rules.
The rules that apply to 78 are the 2, 3, and 6 rules.
The rule that applies to 393 is the 3 rule.
The rules that apply to 76,100 are the 2, 5, and 10 rules.
The rules that apply to 666 are the 2, 3, and 6 rules.

PART C—Activity 1
Answers will vary according to how students group the objects. No object can be used more than once. There should be between 6 and 10 groups.

PART C—Activity 2
Answers will vary according to how students group the shapes. No shape can be used more than once. The students should consolidate their groupings from Activity 1 into three groups.

Lesson 81
PART A

1. 100
2. 42
3. 9
4. 30
5. 3
6. 35
7. 40
8. 9
9. 30

PART B
Students' trees may differ depending on the facts they used, but the prime factors will be the same.

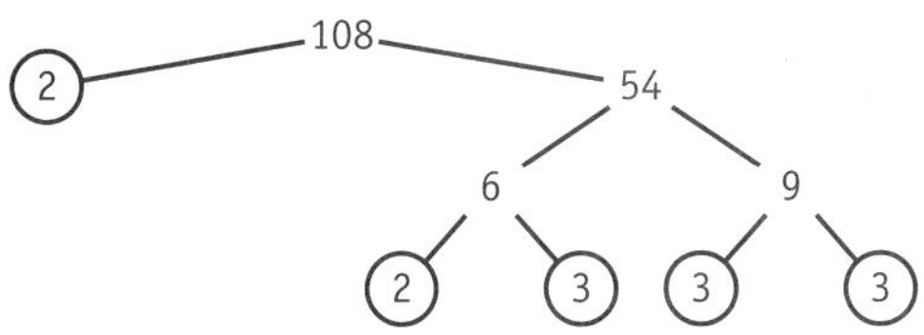

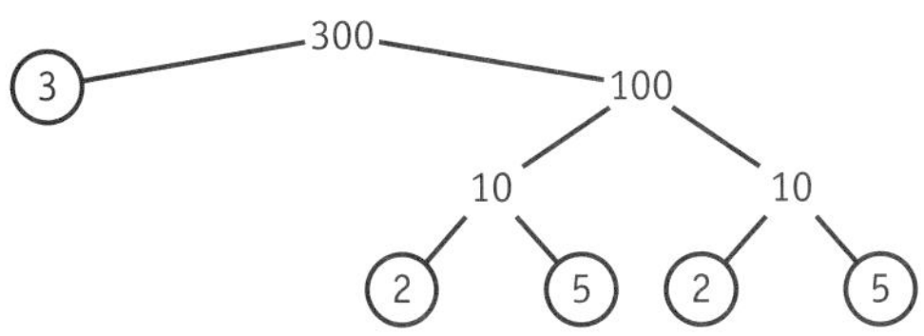

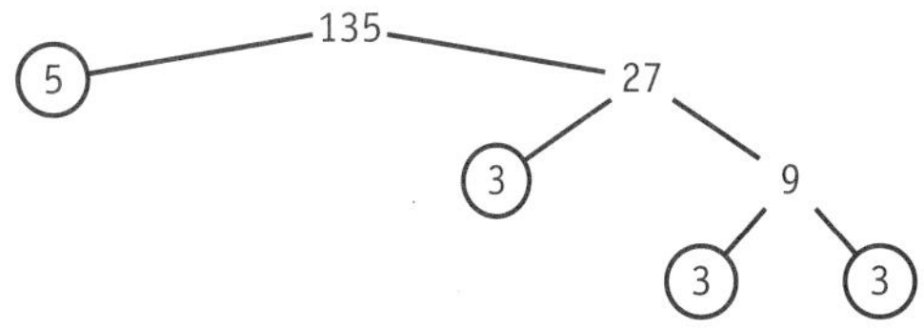

PART C
The answers will vary. The students should use the same shape more than once most of the time. The ideal answer uses many of the shapes several times. For example, all of the shapes have lines. All of the shapes but 1 and 4 are "flat," or two-dimensional. Shapes 5 and 11 are star shapes, but they can be used in a larger group of shapes with straight lines only.

Lesson 82
PART A

1. The factors of 16 are 1, 2, 4, 8, and 16.
2. The factors of 21 are 1, 3, 7, and 21.
3. The factors of 32 are 1, 2, 4, 8, 16, and 32.
4. The factors of 49 are 1, 7, and 49.
5. The factors of 56 are 1, 2, 4, 7, 8, 14, 28, and 56.

PART B

1. The common factors of 12 and 15 are 1 and 3.
2. The common factors of 8 and 12 are 1, 2, and 4.
3. The common factors of 7 and 14 are 1 and 7.

PART C
There are 30 squares in the section of the checkerboard that is circled.

See individual student responses with regard to how they got their answers. Look for this type of discussion:

I saw a big square, so that's one. It's 4×4. Then I saw there were 4 squares that were 3×3. And there were 9 squares that were 2×2. And then there were 16 squares that were 1×1. I added all of those up. $1 + 4 + 9 + 16 = 30$.

Lesson 83
PART A

1. The factors of 20 are 1, 2, 4, 5, 10, and 20.
2. The factors of 18 are 1, 2, 3, 6, 9, and 18.
3. The factors of 28 are 1, 2, 4, 7, 14, and 28.

4. The factors of 42 are 1, 2, 3, 6, 7, 14, 21, and 42.
5. The factors of 40 are 1, 2, 4, 5, 8, 10, 20, and 40.

PART B
1. The common factors of 15 and 20 are 1 and 5.
2. The common factors of 18 and 22 are 1 and 2.
3. The common factors of 10 and 30 are 1, 2, 5, and 10.

Lesson 84
PART A
1. The factors of 6 are 1, 2, 3, and 6.
2. The factors of 18 are 1, 2, 3, 6, 9, and 18.
3. The factors of 25 are 1, 5, and 25.
4. The factors of 32 are 1, 2, 4, 8, 16, and 32.

PART B
1. The GCF of 14 and 21 is 7.
2. The GCF of 5 and 20 is 5.
3. The GCF of 14 and 18 is 2.
4. The GCF of 8 and 12 is 4.

Lesson 85
PART A
1. The factors of 9 are 1, 3, and 9.
2. The factors of 12 are 1, 2, 3, 4, 6, and 12.
3. The factors of 16 are 1, 2, 4, 8, and 16.
4. The factors of 25 are 1, 5, and 25.

PART B
1. The GCF of 10 and 25 is 5.
2. The GCF of 6 and 18 is 6.
3. The GCF of 16 and 24 is 8.

PART C
The rectangle is the shape that makes up Design A. The triangle is the shape that makes up Design B. The parallelogram is the shape that makes up Design C.

Lesson 86
PART A
1. The factors of 30 are 1, 2, 3, 5, 6, 10, 15, and 30.
2. The factors of 35 are 1, 5, 7, and 35.
3. The factors of 27 are 1, 3, 9, and 27.
4. The factors of 22 are 1, 2, 11, and 22.

PART B
Students' trees may differ depending on the facts they used, but the prime factors will be the same.

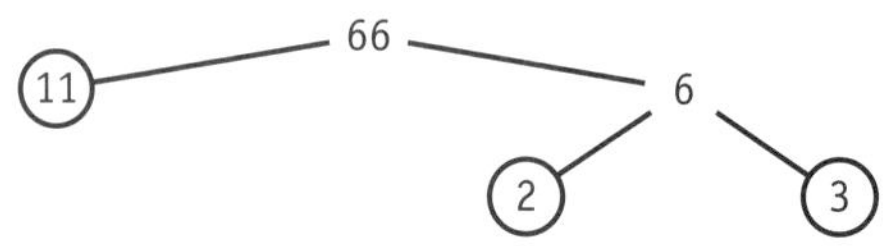

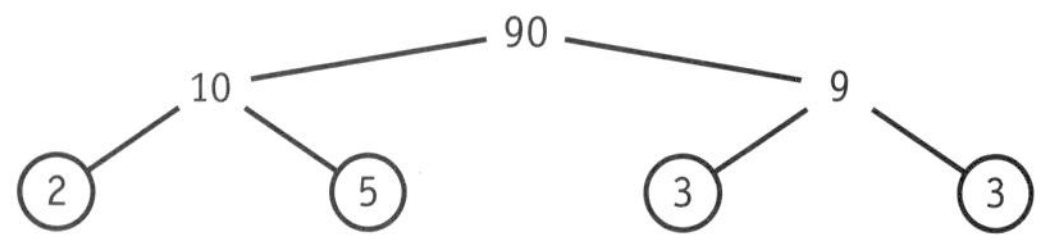

PART C
See individual student work. Look for at least five different shapes made from the triangles. Be sure students have drawn the triangles so that at least one side is lying flat against a side of another triangle.

Lesson 87
PART A
1. 10
2. 81
3. 7
4. 5
5. 4
6. 64
7. 5
8. 9
9. 49

PART B
Students' trees may differ depending on the facts they used, but the prime factors will be the same.

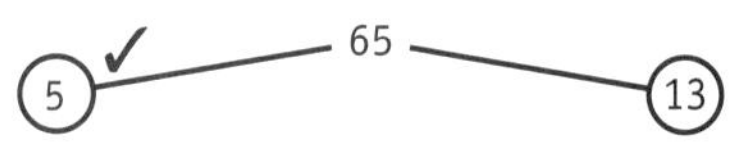

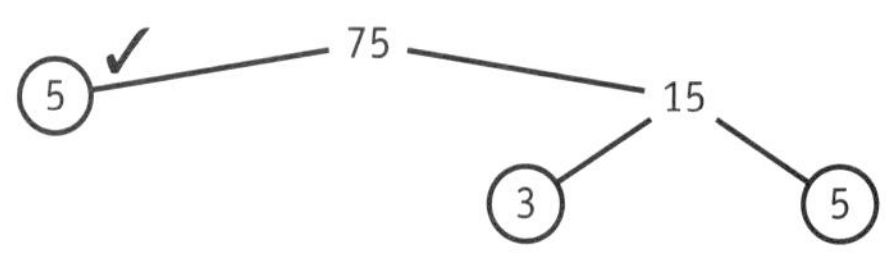

The GCF is 5.

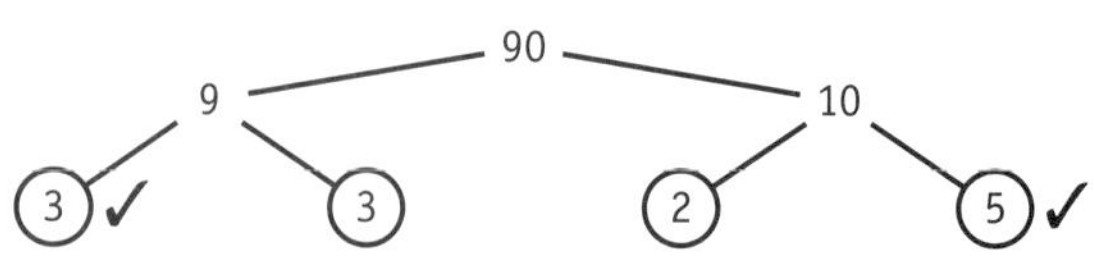

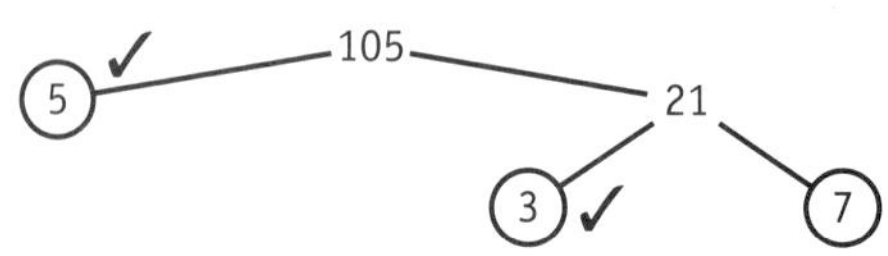

The GCF is $3 \times 5 = 15$.

Lesson 87
PART C
Here are sample answers. Other answers are acceptable. Be sure to check for congruent triangles.

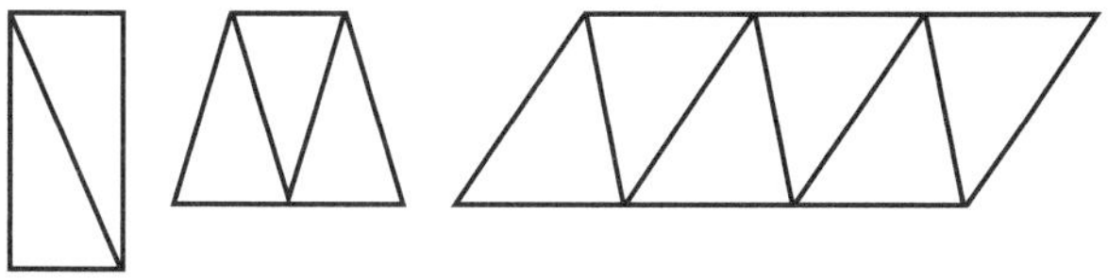

Lesson 88
PART A
1. The rule for 475 is the 5 rule.
2. The rule for 76 is the 2 rule.
3. The rule for 783 is the 3 rule.
4. The rules for 42,010 are the 2 rule, the 5 rule, and the 10 rule.
5. The rules for 822 are the 2 rule, the 3 rule, and the 6 rule.

PART B
1. The area of Object A is 4 sq. units.
2. The area of Object B is $\frac{1}{2}$ of $6 \times 6 = \frac{1}{2}$ of $36 = 18$ sq. units.
3. The area of Object C is $\frac{1}{2}$ of $4 \times 4 = \frac{1}{2}$ of $16 = 8$ sq. units.

PART C
Students' trees may differ depending on the facts they used, but the prime factors will be the same.

1.

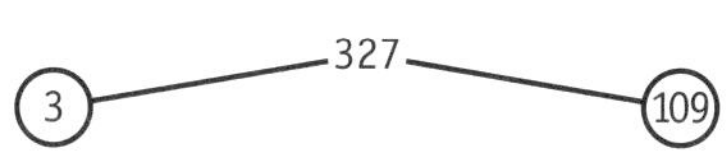

2.

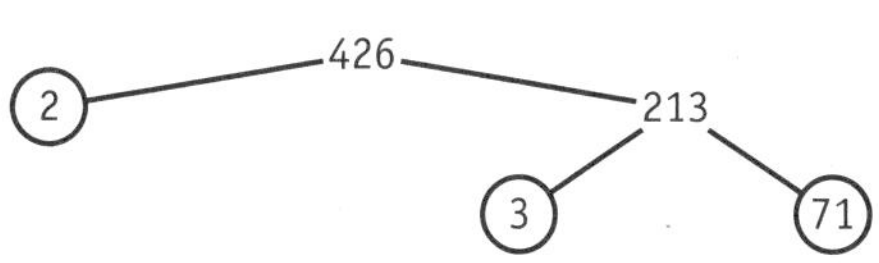

3.

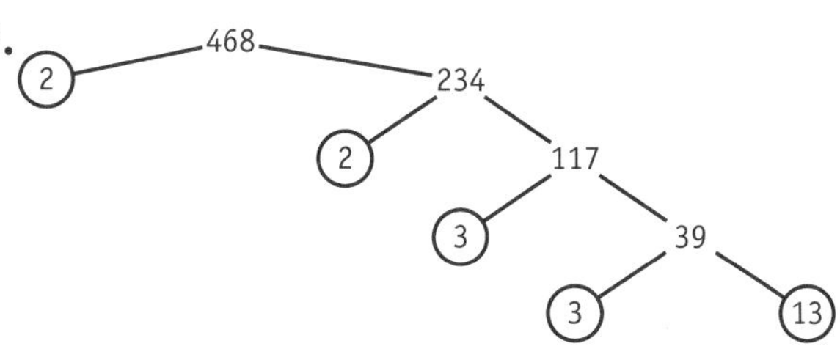

PART D
1. The common factors of 5 and 20 are 1 and 5.
2. The common factors of 7 and 21 are 1 and 7.
3. The common factors of 8 and 24 are 1, 2, 4, and 8.

PART E
1. The GCF of 12 and 24 is 12.
2. The GCF of 16 and 32 is 16.
3. The GCF of 20 and 40 is 20.

PART F
The shapes that are congruent are triangles E, G, and H.

PART G
The GCF of 35 and 75 is 5.

Lesson 89
PART A
1. 110
2. 1,500
3. 210
4. 60
5. 500
6. 100

PART B
(Item in bold type is given in the text.)

Problem	The Answer Will Be: Even Number	Odd Number
6,349 + 76,945	**X**	
987 + 1,854		X
889 × 935		X
93,456 + 123,554	X	
773 + 595	X	
87 × 359		X
328,340 + 222	X	
132,602 × 67	X	
22,421 + 648,448		X
393 × 107		X

PART C—Activity 1

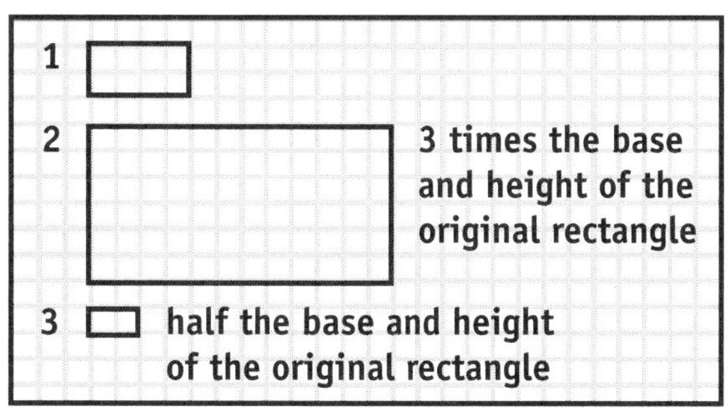

PART C—Activity 2

Rectangle	Perimeter	Area
1	12 cm	8 sq. cm
2	36 cm	72 sq. cm
3	6 cm	2 sq. cm

1. It gets bigger.
2. It gets bigger.
3. 70 sq. cm or 36 times bigger.

Lesson 90
PART A

1. 81	**4.** 49	**7.** 25
2. 640	**5.** 360	**8.** 160
3. 900	**6.** 8,100	**9.** 4,900

PART B—Activity 1

See individual student work. Be sure that there are 10 squares drawn and that they are each 1 cm bigger in base and height than the previous square.

PART B—Activity 2

(Items in bold type are given in the text.)

Size of Square	Area	Difference Between Two Squares
1 × 1	**1 sq. cm**	
2 × 2	**4 sq. cm**	**3 sq. cm**
3 × 3	9 sq. cm	5 sq. cm
4 × 4	16 sq. cm	7 sq. cm
5 × 5	25 sq. cm	9 sq. cm
6 × 6	36 sq. cm	11 sq. cm
7 × 7	49 sq. cm	13 sq. cm
8 × 8	64 sq. cm	15 sq. cm
9 × 9	81 sq. cm	17 sq. cm
10 × 10	100 sq. cm	19 sq. cm

Look at individual student work for answers to the questions. The answers should include information like that indicated here.

1. What patterns do you see in the table and organized list that you have created?
   Look for an answer that includes some of the following discussion: The areas are all square numbers. The difference between the square numbers are ascending consecutive odd numbers (in simpler terms).
2. What would be the difference in area between the next two consecutive squares (that is, between 10 × 10 and 11 × 11)?
   The difference would be 21, because that's the next odd number.

Lesson 91
PART A

1. The factor rainbow for 25 is:

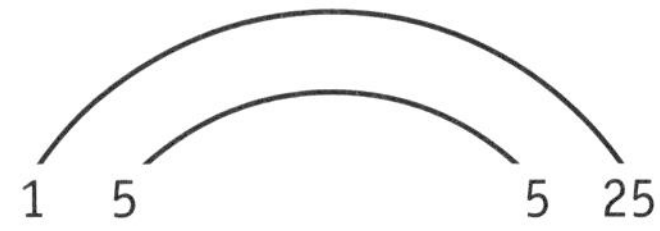

The factors of 25 are 1, 5, and 25.

2. The factor rainbow for 32 is:

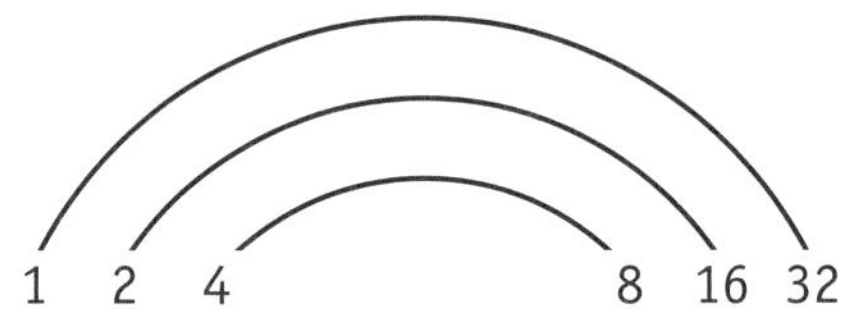

The factors of 32 are 1, 2, 4, 8, 16, and 32.

3. The factor rainbow for 23 is:

The factors of 23 are 1 and 23.

PART B

(Items in bold type are given in the text.)

Look at individual student work. Be sure there are as many square blocks on the page as possibly will fit. The list of number of blocks added should be a list of consecutive ascending odd numbers, starting with 3.

Number the Array Represents	Description of the Array	Number of Blocks in the Array	Number of Blocks You Added
1	**1 × 1**	**1**	
4	**2 × 2**	**4**	**3**
9	3 × 3	9	5
16	4 × 4	16	7
25	5 × 5	25	9
36	6 × 6	36	11
49	7 × 7	49	13
64	8 × 8	64	15
81	9 × 9	81	17
100	10 × 10	100	19
121	11 × 11	121	21
144	12 × 12	144	23
169	13 × 13	169	25

PART C

See individual student work. Look for a triangle with a base of 15 and a height of 8.

The area of the original triangle is $\frac{1}{2}$ of 5×4, or $\frac{1}{2}$ of 20, or 10 sq. cm.

The area of the new triangle is $\frac{1}{2}$ of 15×8, or $\frac{1}{2}$ of 120, or 60 sq. cm.

Lesson 92

PART A

1. 3 **2.** 15 **3.** 36 **4.** 6 **5.** 21 **6.** 45 **7.** 10 **8.** 28 **9.** 55

PART B

(Items in bold type are given in the text.)

Answers From Part A	8 Times Your Answer Plus 1	Check to see if you're right
1. 3	**3 × 8 + 1 = 25**	**5 × 5 = 25**
2. 15	15 × 8 + 1 = 121	11 × 11 = 121
3. 36	36 × 8 + 1 = 289	17 × 17 = 289
4. 6	6 × 8 + 1 = 49	7 × 7 = 49
5. 21	21 × 8 + 1 = 169	13 × 13 = 169
6. 45	45 × 8 + 1 = 361	19 × 19 = 361
7. 10	10 × 8 + 1 = 81	9 × 9 = 81
8. 28	28 × 8 + 1 = 225	15 × 15 = 225
9. 55	55 × 8 + 1 = 441	21 × 21 = 441

Lesson 93

PART A

Students' trees may differ depending on the facts they used, but the prime factors will be the same.

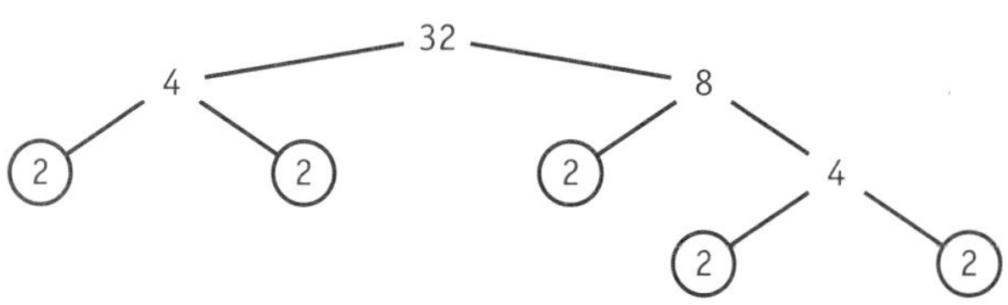

The prime factorization is $2 \times 2 \times 2 \times 2 \times 2$.

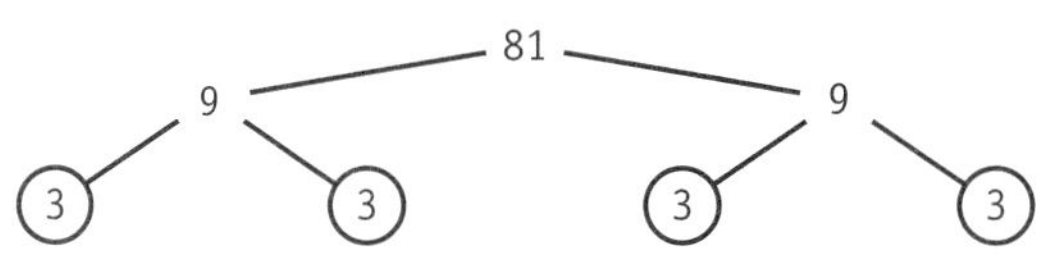

The prime factorization is $3 \times 3 \times 3 \times 3$.

PART B—Activity 1

1. 5^4
2. $2^3 \times 3 \times 5$
3. $2^2 \times 3^2 \times 5 \times 7$
4. 10^9
5. $3^2 \times 7$

PART B—Activity 2

1. 3×3
2. $7 \times 7 \times 7 \times 7$
3. $9 \times 9 \times 9$
4. $2 \times 2 \times 3 \times 3$
5. $5 \times 5 \times 5 \times 4 \times 4$

PART C

See individual student work. Look for a design that is 2 times as big as the original. That means that the trapezoid should have a base of 12 and a height of 4. The triangle above the trapezoid should have a base of 4 and a height of 4. The two side triangles should each have a base of 4 and a height of 4.

Lesson 94

PART A—Activity 1

1. 4^5
2. 30^5
3. $8^2 \times 3^2$
4. $10^3 \times 9^6$
5. 6×8^2

PART B—Activity 1

1. 7: 7, 14, 21, 28, 35
2. 9: 9, 18, 27, 36, 45
3. 5: 5, 10, 15, 20, 25
4. 2: 2, 4, 6, 8, 10
5. 10: 10, 20, 30, 40, 50

PART B—Activity 2

1. The only common multiple shown for 5 and 7 is 35.
2. The common multiples shown for 2 and 3 are 6, 12, 18, and 24.
3. The common multiples shown for 5 and 10 are 10, 20, 30, and 40.
4. The only common multiple shown for 6 and 8 is 24.
5. The only common multiple shown for 7 and 3 is 21.

PART C—Activity 1

1. Turn
2. Slide
3. Turn
4. Flip
5. Turn
6. Slide
7. Turn

PART C—Activity 2

See individual student work. Be sure the students have shown a slide, a flip, and a turn for each object.

Lesson 95

PART A—Activity 1

1. $5^2 \times 3^3$
2. $6^4 \times 4$
3. 7×9^4
4. $8^3 \times 7^3 \times 6^3$
5. $3^2 \times 100^2$

PART A—Activity 2

1. 5: 5, 10, 15, 20, 25
2. 3: 3, 6, 9, 12, 15
3. 8: 8, 16, 24, 32, 40
4. 9: 9, 18, 27, 36, 45
5. 4: 4, 8, 12, 16, 20

PART B

1. The LCM of 5 and 10 is 10.
2. The LCM of 2 and 7 is 14.
3. The LCM of 4 and 6 is 12.
4. The LCM of 6 and 8 is 24.
5. The LCM of 3 and 4 is 12.

PART C

Objects with lines that show symmetry: A, D, E
Objects without lines that do not show symmetry: B, C, D, F
Check student answers to see that they have redrawn B, C, D, and G so that they have lines of symmetry.

Lesson 96

PART A—Activity 1

Students' trees may differ depending on the facts they used, but the prime factors will be the same.

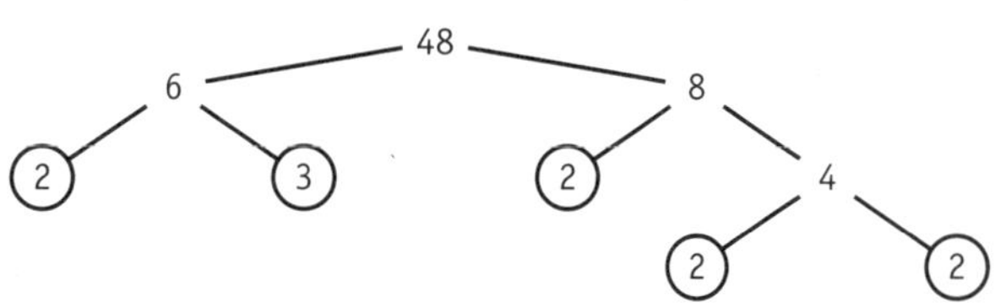

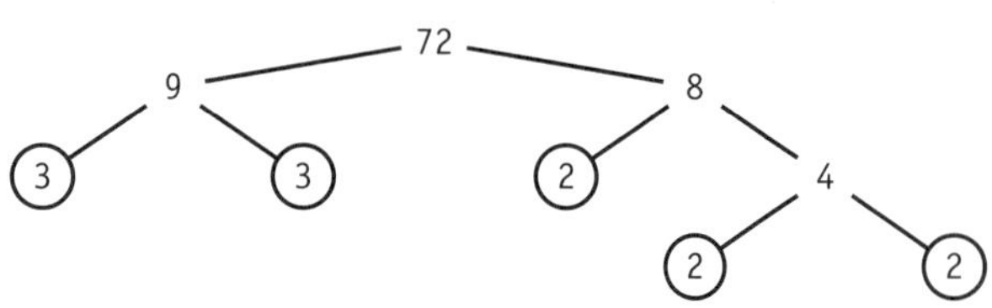

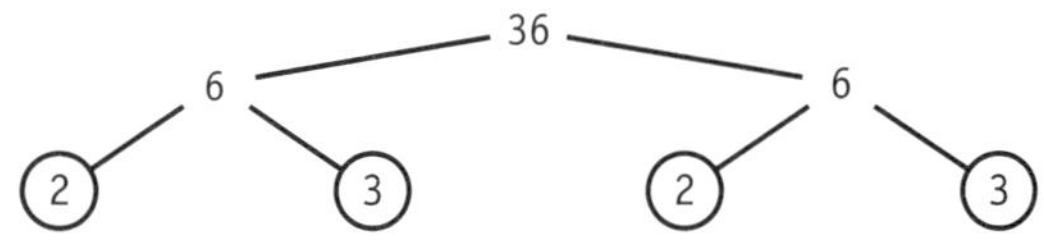

Lesson 96

PART A—Activity 2

1. 4: 4, 8, 12, 16, 20
2. 10: 10, 20, 30, 40, 50
3. 7: 7, 14, 21, 28, 35
4. 5: 5, 10, 15, 20, 25
5. 3: 3, 6, 9, 12, 15

PART B

1. LCM is 18
2. LCM is 28
3. LCM is 40
4. LCM is 30

PART C

See individual student work. Be sure the lines they draw divide the object in half so that the two sides are symmetrical—that is, one side is the mirror image of the other.

Lesson 97

PART A

1. 4^5
2. $3^4 \times 5$
3. $2^2 \times 9^2 \times 10$
4. $5^2 \times 4^2 \times 3^2 \times 2^2 \times 1^2$
5. $1{,}450^4$

PART B

1. The LCM is 8.
2. The LCM is 21.
3. The LCM is 30.
4. The LCM is 15.

PART C

See individual student work. Look for one or more types of objects. Look for slides, flips, or turns. Also, look to see if the mobile has symmetry.

Lesson 98

PART A

See individual student work. Check the following:

1. Make sure these are prime. Examples: 3, 5, 7, 11, 23
2. Make sure these are composite. Examples: 4, 6, 12, 24, 36
3. Make sure these are square numbers. Examples: 1, 4, 16, 25, 36, 49

4. Make sure these are powers of 10. Examples: 10, 20, 30, 40
5. Make sure these are numbers that are divisible by 6. Examples: 96, 18, 72

PART B
1. LCM is 10
2. LCM is 12
3. LCM is 14

PART C—Activity 1

PART C—Activity 2
See individual student work. Look for rotational symmetry with the shape they have drawn; that is, no matter which way you spin the object around the center, it looks the same.

Lesson 99
PART A
1. Yes
2. Yes
3. No
4. No
5. Yes
6. No
7. Yes
8. Yes
9. Yes
10. Yes

PART B
See individual student work. Look for a variety of answers that use the concepts learned in Unit 5.

PART C
See individual student work.

Lesson 100
PART A
1. Even
2. Even
3. Odd
4. Odd

PART B
1. Even
2. Odd
3. Even
4. Even

PART C
1. 25
2. 16
3. 49
4. 81

PART D
The triangular number 6 would look like this:

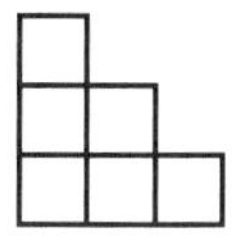

The triangular number 21 would look like this:

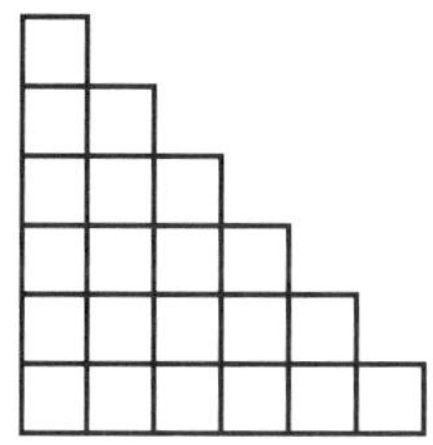

The triangular number 28 would look like this:

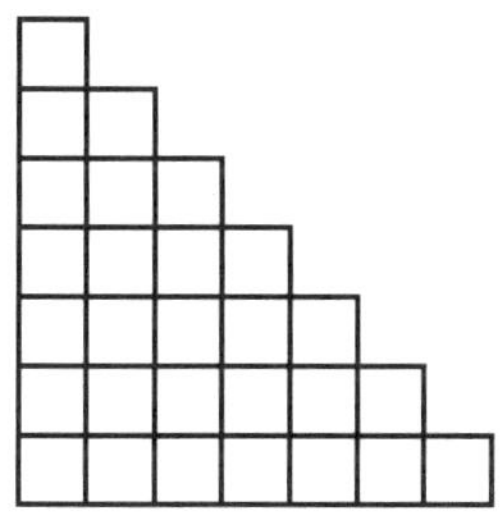

PART E
1. 7^3
2. 3^2
3. 5^4
4. 10^{11}
5. 2^3

PART F
1. The common multiples shown for 5 and 15 are 15, 30, and 45.
2. The common multiples shown for 4 and 6 are 12, 24, and 36.

PART G
1. Slide
2. Flip

PART H
1. The LCM of 3 and 6 is 6.
2. The LCM of 2, 4, and 6 is 12.

PART I
1. Yes
2. No
3. No
4. Yes
5. Yes